STOCHASTIC PHILOSOPHY OF COSMOGENESIS

GENERAL THEORY OF CREATION AND EVOLUTION

MICHEL XILINAS

STOCHASTIC PHILOSOPHY OF COSMOGENESIS

GENERAL THEORY OF CREATION AND EVOLUTION

EDITION NOVEMBER 2013

ISBN 978-1-304-54846-7

CONTENTS

INTRODUCTION

The Stochastic Philosophy of Cosmogenesis is the revised version of my earlier book "Philosophy of the Universe"[1] that includes specific answers to questions that were raised on the matters dealt earlier and required direct clarification. The title of stochastic is defined from the Greek meaning of the word στοχαστικός that is capable of aiming the target (στόχος). This revised version deals on all the questions asked in discussions or comments by many readers and reviewers in the past few months.

Much earlier, in my first philosophical book[2], I had incorporated thoughts, observations and notes that I compiled in margin of my professional career. The guiding principles that I had discussed were:

a) The tetractys or foursome information structure in the evolution of the Universe, where the Earth is, including living organisms

b) The information structure, the anatomy and physiology of the mind,

c) The relationship between material and immaterial information notions following the principles of Parmenides of "being"

d) The ethical determinism of biology

In this book I discuss subjects related to the genesis of the Universe concluding to a general theory of creation and evolution. In this approach, experimental evidence and scientific hypotheses are presented leading to the conclusion of an automatic endless creation prior to the initial Bing Bang up to the development of the human mind.

The basis of the automatic pre Big Bang creation is the postulation of a situation defined as "beyond" or επέκεινα, that is free of all dimensions, time, fields, functions, variables and other properties, excluding the function of bouncing from virtual particles to real particles. The Casimir

---

[1] Philosophy of the Universe – Creation and Evolution. Lulu Press. 2011

[2] XILINAS M. E (Ξυλινά, M.E.) Presocratic Philosophy and Contemporary Physical Sciences, Dodoni, 1997.

effect and Heisenberg's uncertainty principle are used as supportive evidence.

PART 1. NATURALIS presents the natural evolution since the Big Bang describing the fundamental interactions, the general purpose and the variables influencing the cosmogony until the emergence of the human mind.

PART 2 GENESIS OF THE UNIVERSE compiles the natural evolution on earth ending in the emergence of man who is projected in the material environment and is differentiated by his mind functions. The natural history is historically compiled focusing in the development of man and linking the inanimate evolution to the appearance of living organisms based on biology and genetics.

PART 3. METAPHYSICS analyzes the notions as an information system and a revision of the classical Psychobiology concepts, focusing on the mind in the evolution of the Universe. I discuss idealism and materialism and propose a novel philosophical model for ideas, the mind and the human societies positioned in the evolutionary process of the Universe.

Part 4. GENERAL THEORY OF EVOLUTION calls to the cosmogenesis of the Universe from the creation and onwards to the biosphere and the human mind, describing the physical process for the foundation of the Universe. Proposition is made for a spontaneous creation model based on a virtual particle creating an initial scalar field that leads to the creation of Universes without prior requirement or cause.

The twenty-nine plates illustrate the principles described in the various chapters. A chapter resuming the main points of the book follows as QUESTIONS AND ANSWERS.

Greek was the original text and I have translated it in English and is self-published; I apologize for all the numerous linguistic and other discrepancies.

This book is dedicated to my wife Bernadette grateful for her unrestricted support and patience and to the memory of my late sister.

Michel Xilinas

November 2013

PART 1 NATURALIS

In the detectable Universe,[3] spatial inflation (explosive expansion) is taking place initiated from a minute in dimensions point that continues to expand thirteen billion years later. The initial high temperatures while dropping in time allowed the nucleosynthesis from elementary particles and the formation of atoms, molecules up-to the appearance of living organisms. These phenomena are observable and scientifically proven as to the Earth and the surrounding Universe. The contained energy in this vast created Universe is constant under stable pressure and the variables during the elapsed time are the increase in the volume of the space where the explosive inflation is occurring and the drop of the temperature.

Plate 1 presents the vector field of the evolution of the Universe since the Big Bang[4] to date, axis χ showing time and axis ψ^1 temperature. In parallel in this span, the drop of temperature occurs: the temperature in the first millionths of seconds was ten billion Kelvin (K) degrees when now it is only three. In the same scheme, the Plate shows the corresponding logarithmic explosive inflation of the Universe (axis χ showing time and axis ψ^1 temperature and ψ^2 inflation). The dimension to the forefront of the expanding Universe is about 4.6×10^{27} kilometers using today's astronomical capacities.

---

[3] In physics, space-time is any mathematical model that combines space and time into a single continuum. The interpretation of space-time is usually with space being three-dimensional and time playing the role of a fourth dimension that is of a different sort from the spatial dimensions. According to the Euclidean space perceptions, the Universe has three dimensions of space and one dimension of time.

[4] The Big Bang is the prevailing cosmological theory of the early development of the Universe. The theory is supported by the most comprehensive and accurate explanations from current scientific evidence and observation. According to the best available measurements as of 2010, the initial conditions occurred about 13.3 to 13.9 billion years ago. (1.33 to 1.39×10^{10}). Until recently, we thought that the Universe is closed and its dimensions do not surpass 10^{26} meters and in fewer than 10^{11} years, it will implode and disappear. Today cosmologists believe that the Universe will never end and it will continuously generate peripheral daughter Universes (multiverse).
http://www.creationofUniverse.com/html/creation_atom02.html

The space that contains scattered amalgams of material entities[5] celestial corpses (stars, planets, comets, nebulae, galaxies, quasars and energy as radiation), elementary particles, dark matter and dark energy, is described as the Universe. The result of the inflation and unfolding of the Universe is the actual Universe where Earth and the Humans are.

DEFINITIONS

- Laws of Science (laws of Physics)[6] constitute the rules of all actions in the Universe in the entire space-time since the Big Bang.

- The governance of all actions is by the occurrence of either determinism or indeterminism of randomness.

- Entities[7] (Οντότητες) include all the provably existing material and the immaterial.

---

[5] Corporeal, with material form entity (body or substance) relating to, or composed of matter (in contrast to immaterial or incorporeal).

[6] The laws of science (laws of physics including Newton's Three Laws of Motion, Newton's Law of Gravity, Einstein's Theory of Relativity, Quantum Physics, Energy Conservation Laws of Physics, Laws of Thermodynamics and quantum mechanics. These are established scientific laws considered universal and invariable facts of the physical world to date. However if new experimental or scientific evidence is described or discovered then automatically the list of laws of science is adjusted to the new facts. Such laws are: Archimedes' principle, Kepler's three laws of planetary motion, Newton's three laws of motion, Euler's laws of rigid body motion, Newton's law of universal gravitation, heat, energy, and temperature, Newton's law of cooling, Boyle's law, law of conservation of energy, Joule's first and second law, the four laws of thermodynamics, quantum mechanics, Heisenberg's uncertainty principle. Other relevant laws are the chemical laws, Ohm's law, Gauss's law for magnetism, the thermodynamic laws, Schrödinger equation, the radiation laws, Planck's law of blackbody, and the laws of Kepler among others.

[7] Adapted from Aristotle :
- A thing's material cause is the material of which it consists.
- A thing's formal cause is its form.
- A thing's efficient or moving cause is the primary source of the change.
- A thing's final cause is its aim or purpose.
By adapting the Aristotelian definitions the entities, here, as causes are the following:
a. A material entity's cause is the nature out of which it consists: in the Universe, it is its energy, sub-atomic, atomic and molecular structure; for each immaterial it is the abstract idea that describes it i.e. number, formula, form..
b. For a material object, its formal cause is the numerical (digital) abstract information that fully describes it and from which it is exactly reproducible.

- Materials are all energy, radiation or particle containing entities that are provably existing and are real.

- The Universe contains[8] energy including radiation and particles and other equivalent forms of energy.

- Immaterial entities are abstract notions containing concrete information. All immaterial are free of matter and energy and are independent of all properties including space-time, variables or other dependencies.

- Movement is the displacement of a material entity under the influence of a momentum between two-time points. The movement ends to an achieved action (πρᾶξις, praxis) event, which is registered as a snapshot.

- A snapshot is an exact and precise descriptive depict of a material entity of all properties and space-time relations as to all other existing entities and includes its variables and dependencies that allows its replication.

- Action is the force that causes movement of one or more entities that constitutes an event and a respective snapshot.

- Potential action (δράσις) is a subjective tactical approach towards achieving an event but at the preparative and planning process and is one among other possible and potential actions concerning the precise context.

- Achieved action (πρᾶξις) is the final implementation among the potential actions that concludes a movement or an event, and is registered by a concrete snapshot.

- State is the description of the properties of any material or immaterial entity defining circumstances, conditions, status, context or compatibility.

- Event is any movement or change that results from an action in any material entity in the space-time. Each such event is registered and classified by two concrete snapshots that depict the snapshots of the entities moved within the total context of all the other materials in the Universe, at the concrete instance in time.

---

c. An entity's action cause is the force (dynamis, δύναμις) that will power its potential or accomplished action.

[8] The known to science radiation and elementary particles are the materials that describe the Universe. In general, all the particles are fermions or gauge bosons; the fermions comprise the electrons, protons, neutrons, quarks and neutrinos. The elementary particles that form the materials are fermions that are either quarks (protons and neutrons) or leptons (electrons and neutrinos) that make-up matter that build the celestial bodies and in the biosphere living organisms including human beings and their constructions.

- Momentum is the product of mass multiplied by the speed and is the force that potentially can end-up in the implementation of an action by a material entity[9] under the influence of a force.

- Purpose is the theoretical overall probabilistic strategy towards an action by taking in context objectively the totality of similar states.

- Will is the force that powers all movements and changes of material entities and is either active or passive.

- Governance in the Universe is the scientifically proven framework that commands and manages the administration of the Universe. Governance is deterministic, indeterministic, or random. All actions in the Universe are processed through governance.

- Evolution is the natural history of all registered snapshots from the first creation of the Universe to a time point. Evolution is thus the accomplishment of all the actions as depicted by the respective snapshots and processed by the governance.

- The governance rules based on the laws of Science the evolution in the Universe since creation; the fundamental interactions power it.

- Humans taxonomically *Homo sapiens* have survived as the only species of the *Homo* genus.

- Ascending pathway (trend) of evolution in the Universe since creation is the subjective descriptive historical observation that leads to the mind of Humans.

- Noocentric evolution is the pathway followed in the natural history from creation of the Universe to date, which among an unlimited number of potential evolutionary pathways finally led to human mind.

NATURAL NOOCENTRIC EVOLUTION

The combination of the concurrently occurring expansive inflation of the Universe to the drop of temperature summarizes the apparent ascending evolution. Plate 2 presents the stages of the evolution since the Big Bang (axis χ showing time, axis Ψ^1 temperature, axis Ψ^2 expansive inflation and axis z the evolution in the Universe). The Plate presents the successively

---

[9] The entities are real and existing following the Parmenidean principle of "εἶναι" and therefore are provably detected and observed in the Universe. Materials (things, matter) are entities made up from energy or particles that "εἶναι" that really exist in the space-time of the Universe obeying to the laws of science. Mathematical, energy, chemical or wave equations are abstract immaterial notions that describe their respective material entities.

various stages that are the pivotal echelon milestones in the ascending track of the Universe toward the human societies of today.

Already at the first instances of the evolution, when high temperatures pertain (quarks and nucleosynthesis) up to the synthesis of the DNA and RNA[10] of nucleotides and the appearance of the first living microbes, it is noted that a series of a chain of interdependent and linked sequences of tetractides are deployed (Plate 2).

Tetractys is the first number formed by the addition and multiplication of equals; for the Pythagoreans it was representing justice as the first number that is divisible in every way into equal parts. Empedocles[11] the philosopher from Agrigento in Sicily described the same tetractys as constituting the basic elements of all living organisms. The tetractys symbolized in the Greco-Roman era, the foursome base elements, earth, air, fire, and water, thought as the roots of the Universe. In this book we discuss our observation that in the evolution follows a sequential series of tetractides as observed by extrapolating the Pythagorean, and the later Empedoclean tetractys to the present level of science (Plate 2).

These sequential tetractides are thus introduced as part of the standard model[12] and of the specific noocentric evolution of the Universe.

Each sequential stage of the noocentric evolution is dependent and marked by a set of four-key radicals, the tetractides[13]. These sequential

---

10 DNA, RNA are polymeric constituents of all living cells and many viruses, consisting of a long, usually single-stranded chain of alternating phosphate and ribose units with the bases adenine, guanine, cytosine, and thymine or uracil bonded to a sugar.

11 Empedocles established four ultimate elements, which make all the structures in the world, fire, air, water, earth. Empedocles called these four elements "roots", which, in typical fashion, he also identified with the mythical names of Zeus, Hera, Nestis, and Aidoneus. Empedocles never used the term element (Greek: στοιχεῖον) which seems to have been first used by Plato. The development of the different structures occurs following the proportions in which these four indestructible and unchangeable elements combine with each other. It is in the aggregation and segregation of elements thus arising, that Empedocles like the atomists, found the real process that corresponds to what is popularly termed growth, increase or decrease. Nothing new comes or can arise; the only change that can occur is a change in the juxtaposition of element with element. This theory of the four elements became the standard dogma for the next two thousand years.

12 The Standard Model recognizes two types of elementary fermions: quarks and leptons and describes 24 different fermions: 6 quarks and 6 leptons, each with a corresponding anti-particle. Peebles, PJE. 1998. The Standard Cosmological Model. http://www.ipac.caltech.edu/level5/Peebles1/frames.html

13 The tetractys (τετρακτύς, plural tetractides, foursome, or quartet) was the essence of the Pythagoreans. The sum of the first four natural numbers (1+2+3+4=10) describes the

series of tetractides, that are however different at each epoch of the evolution, lead to the current on the Earth appearance of Humans dowered with mind. The last therefore in the evolution tetractide series is the genetic DNA of living organisms that expresses all the information required by a four-nucleotide code.

The consequent interactions in the space-time provoke the endless movement of the material particles in the Universe that in turn changes at each instance the registered snapshots. Even a minor movement - for example of a photon or of an energy quantity of a Planck length, or mass, charge, or temperature - will change the snapshot to the next. It is therefore proposed that the minimal time between two snapshots is at least equal to the Planck's time[14] (5.39124×10^{-44} seconds) or multiples of it.

Each reference to the space-time concerns the liaison between the time and distance from the creation of the Universe point (Big Bang), which is the start of the evolution process. The expression of the time lapse is in multiples of the Planck's time and describes events occurring in the Universe where the Earth is.

The exercise of a force or combination of forces on a material entity accomplishes a movement. In the microscopic as well macroscopic space of the Universe, movement is the cause of transfer of thermal or other energy from the highest to the lower level of energy in the continuous process of expansion. Every material, starting from an elementary particle, is under the influence of forces that allow possible potential actions that result to a movement and an achieved action that finally is pictures and registered as a snapshot.

If the observable most advanced human cultural society is empirically the highest and most complex state in the Universe, then with a phenomenological approach this is the provisional endpoint at the present snapshot of the evolution in the observable Universe. Here, the expression of this is contained in the definition of the noocentric evolution. The grading of the evolution is that which is involved in the geological material (inanimate) as well as that involved in the biological and third that involved in the psychological with the human mind.

---

tetractys. Numerous authors since antiquity to date extrapolate the tetractys in music, poetry, philosophy, medicine and biology.

[14] Time required for light to travel, in a vacuum, a distance of one Planck's length

Teilhard[15] first reached to a parallel conclusion, the noosphere, however linked to immaterial theological conclusions. In this context, the noocentric evolution is strictly material and related to the evolution.

The noocentric, and in general the natural evolution, is compatible with the possibility of concurrent evolutions in many galaxies. It also stipulates that the noocentric evolution is an observation and does not exclude that it leads to a failure. Nor does it exclude that other evolutions in the Universe are comparatively more successful reaching higher than the observable on earth provisional target: the mind of Humans. Further, it does not exclude the creation of other Universes, similar or different from the Universe where the Earth lies.

Tetractys, tetractides (Foursomes or Quartets)

Earlier[16] it was described the crucial dependence role of the evolution on a tetractys of causative elements. A serial of tetractides of these fundamental causative elements is involved already at the first instances just after the Big Bang in the evolution. Plate 2 presents these sequential series of causative tetractides before the nucleosynthesis, at the subatomic, atomic, molecular and genetic biology nucleotide levels. Influenced by the

---

[15] Pierre Teilhard de Chardin was a French philosopher and Jesuit who trained as a paleontologist and geologist and took part in the discovery of both the Piltdown Man and Peking Man. Teilhard conceived the idea of the Omega Point and developed Vladimir Vernadsky's concept of Noosphere. Some of his ideas came into conflict with the Magisterium of the Catholic Church that censured several of his books. Teilhard's primary book, The Phenomenon of Man, set forth a sweeping account of the unfolding of the cosmos. He abandoned traditional interpretations of creation in the book of Genesis in favor of a less strict interpretation. In this theory, developed by Teilhard in *The Future of Man* (1950), the Universe is constantly developing towards higher levels of material complexity and consciousness. Teilhard's theory of evolution states that the Universe can only move in the direction of more complexity and consciousness drawn by a supreme point of complexity and consciousness and postulates the Omega Point as this supreme point of http://www.answers.com/topic/pierre-teilhard-de-chardin

[16] XILINAS M. E (Ξυλινά, M.E.) Presocratic Philosophy and Contemporary Physical Sciences, Dodoni, 1997

Pythagorean oath[17] of the tetractys, Empedocles[18] observed that four causative radical elements (rizomata) form the basis of all existence of plants, animals and Humans[19]. Extrapolated to contemporary physics and biology the identification of the tetractides in the evolutionary process is an astonishing concept, as is the role of the two Empedoclean forces (love and strife) to generate the motion of the elements (rizomata). The sequential series of the tetractides (Plate 2) lead to the mind of Humans, and this is an observable and provable phenomenon.

The further extrapolation of the two forces proposed by Empedocles links us to the contemporary "grand unified theory" model. This separation probably occurred at one point between 10^{-36} seconds after the Big Bang when the strong and electroweak interactions (interactive forces) are unified leaving the gravitation separate.

Plate 3 (up) shows the interchange of three generations of elementary particles from the time of 10^{-44} seconds to 10^2 seconds after the Big Bang. The four fundamental interactions of the Universe power this

---

[17] "By that pure, holy, four lettered name on high, nature's eternal fountain and supply, the parent of all souls that living be, by him, with faith find oath, I swear to thee."

[18] "Empedocles of Akragas, son of Meton, says that there are four elements, fire, air, water, earth; and two dynamic first principles, love and strife; one of these tends to unite, the other to separate. In addition, he speaks as follows: Hear first the four roots of all things, bright Zeus and life-bearing Hera and Aidoneus, and Nestis, who moistens the springs of men with her tears. Now by Zeus he means the seething and the aether, by life-bearing Hera the moist air, and by Aidoneus the earth; and by Nestis, spring of men, he means as it were moist seed and water. Aet. Plac. i. 3; Dox. 287."
Empedocles makes the material elements four: fire and air and water and earth, all of them eternal, and changing in amount and smallness by composition and separation; and the absolute first principles by which these four are set in motion, are Love and Strife; for the elements must continue to be moved in turn, at one time being brought together by Love and at another separated by Strife; so that in his view there are six first principles; for sometimes he gives the active power to Love and Strife, when he says (vv. 67-68):
.. "Now being all united by Love into one, now each borne apart by hatred engendered of Strife;" and again he ranks these as elements along with the four when he says (vv. 77-80):
.. "And at another time it separated so that there were many out of the one; fire and water and earth and boundless height of air, and baneful Strife apart from these, balancing each of them, and Love among them, their equal in length and breadth". Theophr. Phys. Opin. 3; Dox. 478.
[19]Empedocles Fragments and Commentary Arthur Fairbanks, ed. and trans. *The First Philosophers of Greece* (London: K. Paul, Trench, Trubner, 1898), 157-234.
http://history.hanover.edu/texts/presoc/emp.htm#book3

interchange. The sequential evolutionary tetractides described in Plate 2 follow in time those of Plate 3 (up) from 10^{-44} seconds until 10^2 seconds.

The twelve fundamental fermions are divided into three generations[20] of a tetractide of four particles each (Plate 3, down). Six of the particles are quarks. The remaining six are leptons, three of which are neutrinos, and the remaining three are the electron and its two cousins, the muon and the tau. Each generation of particles forms a tetractide, as does also the column of the bosons.

In the same Plate 3 (down, last column) is shown a fourth generation describing the bosons. In the standard model, there have been described five elementary bosons: the four (tetractys) of γ, g, W$\pm$ and Z that occupy the four positions in the IV column of the Plate and the hypothetical for the present Higgs boson.

Again, in Plate 3 (down), the first three columns present the three generations of tetractides. Two leptons and two quarks make each tetractys of each generation. Taking as a basis, the standard model further generations are mathematically impossible.

Since the Big Bang[21] on the pathway to the mind, the sequence of tetractides milestones observed in seconds is as follows:

- First, the early 10^{-34} seconds period with the elementary particle u, d quark, electron, and neutrino

- Second at 10^3 seconds period with free quarks and leptons forming tetractides of two u quarks, one d quark and electron (nucleosynthesis)

- Third at 3×10^{13} seconds or one million years after the Big Bang, with the atom tetractide synthesis with hydrogen, oxygen, nitrogen, carbon (during the era of recombination to form atoms)

---

[20] Each member of each group of higher generation has a greater mass from the corresponding particle of the precedent one (Plate 4, down). For example, the electron has a mass of 0.511 MeV/c2; the muon of the second generation has 106 MeV/c2, and the tauon of the third generation 1,777 MeV/c2 that is nearly double of that of the proton.

[21] The age of the Universe since the Big Bang is 13.8×10^9 years (or 4.2×10^{17} seconds). The Earth solar system formation was 9.1×10^9 years after the Big Bang and of the Earth 9.2×10^9 (meaning that the age of the sun in years is 4.57×10^9, and the age of the Earth is 4.54×10^9.

- Fourth at 1.17×10^{16} seconds or 3.8×10^9 years after the Big Bang, appear the four-nucleotide tetractides of the DNA or RNA and the first living organisms.

Based on the above observations we conclude that globally from the first instances after the Big Bang until now, the specific evolutionary process towards the human mind is echeloned with pivotal tetractides that form a continuous sequence.

Fundamental interactions

In physics, fundamental interactions are the ways by which the simplest particles in the Universe interact. The four known fundamental interactions are electromagnetism, strong interaction, weak interaction (also known as strong and weak nuclear force) and gravitation[22]. The laws of Science are the government in the Universe of all events since the Big Bang including all material entities[23]. These laws are the basis of the four fundamental interactions and follow the formation of the first subatomic particles at specifically appointed times in the immediate aftermath of the Big Bang to form the entire order and system of the Universe. Atoms, that built the material Universe, owe their existence and even distribution across the Universe to the interaction of these interactions that have a distinct intensity and impact. These interactions allow the formation of the material Universe through a distribution of energy. The Plate 3 (up) shows in a time diagram the evolutionary events that occurred since the start of the explosive inflation to now. The start of this time diagram starts from 10^{-43} seconds when the fundamental interactions where in theory a unified force.

In a later chapter[24], is described a novel approach explaining the first causative instance of the Big Bang connected to the expansive inflation of a primary scalar field.

Governance: determinism, indeterminism and randomness

Plate 4 summarizes the existing material and immaterial entities in the Universe, the governance of determinism, indeterminism or randomness

---

[22] Except gravity, these interactions can usually be described in a set of calculation approximation methods known as perturbation theory, as mediated by the exchange of gauge bosons between particles.

[23] Abstract immaterial notions are independent of laws of Science.

[24] See Part 4 GENERAL THEORY OF EVOLUTION

and the four fundamental interactions (forces). The immaterial entities (i.e. ideas) are free of all interaction of forces while all materials including living organisms are directly dependent on their action.

A direct interdependence exists between the expected response and the governance status. Indeterminism is mathematically[25] a nonlinear system; on the contrary, a deterministic system is one described by an operator as linear.

Determinism, indeterminism and randomness together command all outcomes of all actions leading to events registered as snapshots.

Determinism[26] is the philosophical view that every event, is causally determined (completely predictable) by previous events. In deterministic terms the Universe is ruled by causal laws resulting in one possible only state at any time. Determinism proposes that there is an unbroken chain of prior occurrences back to the origin of the Universe. This applies to the acceptance that a material entity either living organism or mechanistic, is not free to act actively but only passively as a sequence of causative law hood of achieved previously actions.

On the contrary, indeterminism accepts that a state has a free Will and can opt for a potential action, aiming toward an accomplished action. The intrusion of the governance by indeterminism is due to the human mind and all active actions (and respective snapshots) are unpredictable and independent of the antecedent one.

By definition here, in the evolution the appearance of the Human mind initiates indeterministic responses that lead to the achievement of active actions, while deterministic responses lead to passive actions. The mind disposes of programs (perception, verification, response) and methods (inspiration, processing, and implementation) to act. These mind actions are under the governance of indeterminism whenever powered by active

---

[25] In mathematics, a nonlinear system is one which does not satisfy the superposition principle, or whose output is not directly proportional to its input. Less technically, a nonlinear system is any problem where the variable(s) to be solved for cannot be written as a linear combination of independent components. A linear system is a mathematical model of a system based on the use of a linear operator. Linear systems typically exhibit features and properties that are much simpler than the general, nonlinear case. http://en.wikipedia.org/wiki/Nonlinear_system

[26] Determinism is commonly understood as the thesis that "the laws that govern the Universe (or a subsystem), together with the appropriate initial conditions, uniquely determine the entire time evolution of the Universe". Indeterminism is the negation of this thesis.

Will or under determinism or randomness on all other circumstances whenever powered by passive Will.

Randomness is involved in each action whenever the five variables interplay or on the quanta level as will be discussed below.

<u>The five variables in the evolution (the 5 π)</u>

In the evolution, of the Universe there is interplay of a process comprising five variables. These variables are multiplicity, complexity, diversity, plasticity and adaptability[27]. Competing conditions during the long development characterize the evolution process with the rapid multiplication of material entities. These entities are either energy or particles or atoms or molecules or later minerals and living organisms. In parallel, the evolution has kept a trend to advance by developing complex chemical entities and living organisms that are enriching the diversity. Altogether, the complexity and diversity are competitive with increased need of resources. This need renders the plasticity necessary, especially in the living organisms with improvement of adaptability to the environment. The five variables together refer in a more complex manner to the term coined to denote the state of the maximum organized centricity as earlier described by Teilhard[28], by interpreting complexity as the axis of evolution of matter.

The five variables together have as an endpoint the compatibility of the individual materials during the evolution. The compatibility of the developmental process in the evolution is decisive and fluctuates in the frame given by the five variables. The evolution is under the power of Will, under the rule of the laws of Science, under the interplay of the five variables and under the governance of determinism, indeterminism and randomness.

---

[27] In Greek translated to: πολλαπλότητα, πολυπλοκότητα, ποικιλότητα, πλαστικότητα, προσαρμοστικότητα, therefore the five (πέντε) π.

[28] Pierre Teilhard de Chardin conceived the idea of the Omega Point. In this theory, the universe is constantly developing towards higher levels of material complexity and consciousness, a theory of evolution called the Law of Complexity/Consciousness. The Omega Point exists as supremely complex and conscious, transcendent and independent of the evolving universe

The dominance, at least on Earth, of Humans at this point raises another condition that appears in the evolution that is the human mind. Concurrently the Will powers the generation towards the change in the space-time by propulsive action and movement that ends with an event and a registered snapshot. Every change in a snapshot in the space-time is the result of the interactive interplay of these five variables.

The particularity of the evolutionary process is that more the complexity is increased less the possibilities of potential actions happen, at least in the visible horizon of science of today. More precisely, in areas like the biosphere on Earth the process of the evolutionary unfolding reveals that the diversity has increased. The synthesis of a variety of chemical molecules allows the harmonic evolution of livings by increasing the plasticity and adaptability. The complexity of the chemical molecules sets restrictions to the potential of multiplicity and to the generation of other compatible living organisms. The environmental frame of the biosphere in turn causes the decrease of diversity and sets additional restrictions.

In the history of the Universe' evolution at the early instances of the Big Bang, there were restrained numbers of entities and therefore the multiplicity was excessive in contrast to the other four variables that had a negligible effect. However, as the time advanced already in the first billionths of second the other variables start taking a significant role in the evolutionary process with up or downward bounces.

A marked multiplicity occurs in the Universe in the first instances after the Big Bang with the dominance of radiation, gluons and other particles that lead to the nucleosynthesis. The multiplicity and variety of these early entities starts increasing the complexity of the newly synthesized particles, atoms and later molecules and in turn increases the diversity. The mind emerges much later in evolution and follows the effects on the five variables concurrently with the ascending trend in the evolution. As to the human mind, more the human societies are ascending more the uniqueness of the respective mind increases with reduction of multiplicity, while all the other four variables increase.

Active and passive Will

The interface of the four fundamental interactions in nature relates directly to the evolutionary history of the Universe, as historically observed and described. The unification of the four forces remains an open subject in Physics. Plate 3 (up) shows the first instances at the start of the expansive inflation of the Big Bang with a hypothetical unification of the four forces.

The open philosophical question is if there is a link between a type of unified force and the expression of the evolution under the governance of determinism, indeterminism or randomness. This is shown in Plate 2 (dotted line curve Z starting from the Big Bang) illustrating the hypothetical unified force in connection to the development in the Universe. The Z curve presents the ascending development of the Universe until today powered by an arbitrary unified force. Here, we describe the arbitrary Z force that powers the evolution as Will. When we regard Humans as the most advanced event in the Universe at this point then empirically we state that the evolution is ascending.

The natural history of the evolution, powered by the Will that bases the empirical observation that from the Big Bang until the human mind of today, there is an ascending trend. This trend (Plate 2) is on empirical observations of the physical Universe and asserts that the evolution is leading to the human mind during the $1,37 \times 10^{10}$ years elapsed.

Randomness governance

The governance of indeterminism or determinism is in most cases in the evolution concurrent with randomness. Randomness entails uncertainty in the actions and supports unpredictability. This is due to the uncertainty principle of Heisenberg or the interplay of the five variables, as described earlier.

The uncertainty principle of Heisenberg is valid whenever quanta mechanics are applicable; therefore, it is pertinent to the sub-atomic elementary particle physics. In all cases where Newtonian mechanics are applicable, the uncertainty principle is negligible.

Will and governance

The Will powers the evolution since the creation of the Universe and is active or passive. The governance of both, active and passive Will is by determinism, indeterminism, randomness or a combination of them. Passive Will is under the governance of both determinism (wherever the classical mechanics are valid) and randomness (wherever the quanta mechanics are valid or by the uncertainty of the interplay of the five variables). Under the governance of determinism, the Will is passive and the resulting action is predictable. On the contrary, if the Will is active, and in this sense voluntary and interventional, then the governance is by indeterminism which is *a priori* unpredictable. In randomness, the resulting action is unpredictable.

- For all actions operated through the mind, the governance is by indeterminism – a prerequisite being that the evolution has reached to the point of the appearance of Humans

- For all actions in the environment of Humans in the Universe the governance is by determinism (wherever Newtonian Physics are valid) and randomness (wherever quanta mechanics are valid)

- For all actions of all material entities including living organisms other then Humans, the governance is again by determinism (wherever Newtonian Physics are valid) and randomness (wherever either uncertainty due to the quanta mechanics or the uncertainty caused by the five variables)

All actions occurring during evolution in the Universe of any material entity there is an enablement of either the uncertainty principle of Heisenberg or of the interplay of the five variables is present. This enablement varies as follows:

- In all cases whenever quanta mechanics the enablement is valid and strong

- In all cases, where atomic or molecular interactions occur there is enablement of the interplay of the five variables while the uncertainty principle of Heisenberg is weak or insignificant.

- The same is in all cases where Humans or other living organisms are involved as individuals

- The same is in all cases where human societies are involved.

Plate 4 presents the entities in the Universe (material and immaterial) and the forces that apply on them leading toward an action (praxis). Determinism and randomness governance are interchanging in the evolution until the appearance of the human mind. It is depicted also the material action process in relation to the four fundamental interactions of material entities, living organisms, Humans and their societies. It is also illustrated the involvement of Will as well as the governance in the process. Outside the material action process lie the immaterial such as ideas.

The differentiating factor between active and passive Will is the occurrence of indeterminism in the governance, after the appearance in the evolution of the human mind[29]. Both certain animals as well as

---

[29] In Chapter "Information Systems in living organisms", the human mind is differentiated from the nervous system of animals.

Humans share the ability to opt among several alternatives; however, the first by using the mind powered by active Will while the latter by using other communication systems or organs. Responses are indeterministic when they involve the mind and deterministic in the other cases.

GENERAL PURPOSE OF THE EVOLUTION

The purpose of evolution is either particular for a concrete material or general encompassing the functional purpose of all the materials and energy entities that exist in the Universe including the human societies.

With the current knowledge, scientific data, natural history and events in the observable space-time the trend of the evolution is parallel to the ascending levels of human societies. Humans and their societies' evolution, based on the upgrading of the mind, refer to the noocentric evolution. This ascending evolutionary pathway is an obvious position based on the study of the natural evolution and the historic observation and it postulates that the purpose is the rising upgrade of human mind and societies to the highest possible level. Further, each human society collectively also shares a purpose determined by its progress and advancement.

Specifically, the purpose of a living organism in the Universe as an individual is its survival and reproduction in a specific environment. The purpose of simple survival and reproduction Humans share with the other living organisms.

If the purpose of the Universe, based on a human-mind oriented observation, is the ascending evolution of human societies this fact is neither the proof nor even the indication that this trend will continue or that there is no other trend possible or finally that there are no other possible ascending evolutionary pathways. It is also questionable if the future human societies, at least on Earth, are on an ascending and not on a destructive and downward trend in the long-term.

SNAPSHOTS

Even the most single and individual event in the Universe constitutes a snapshot depicting the material entity concerned, its properties and its space-time relations as to all other existing entities. Every snapshot in space-time depends on a sequential chain association with the states that occurred just before; no overlaps can occur[30].

---

[30] As long as the are no particles faster then the speed of light (tachyons)

A snapshot reflects the change of each individual material entity in time and therefore snapshots constitute time sequence continuity and not a space location one: by material entity meaning in terms of quanta mechanics either an elementary particle or quanta energy.

The snapshots form a continuous registry from the first instance of the creation of the Universe to the actual time. The absolute division of this registry is in Planck's time intervals, however for practical reasons, a relative time registry is valid by calculating the conventional time since the Big Bang. Each Planck's time interval constitutes a slot during which an action on an entity occurs, creating an event and registered as a snapshot.

If the Universe since the start is 13.8×10^9 years (or 4.2×10^{17} seconds) and Planck's time is roughly 10^{-43} seconds then the registry of the snapshots to date contains approximately 4.2×10^{60} Planck' time.

Primordial monoclonal snapshots

Whenever the sequentiality of the snapshots is continuous in time, we define these snapshots as monoclonal. At the first instance of the creation of the Universe, a first event occurs registered as the primordial snapshot. From this point on a sequential and continuous series of snapshots will follow forming a direct uninterrupted series of the snapshots: we define this first series as the primordial monoclonal.

The governance of this primordial series is by determinism and is predictable.

There is a continuous increase of the number of snapshots in time irrelevant if they are distant in time or in location between them. Each new snapshot in the series is consecutive to its previous with a varying time interval and distance length and the governance is deterministic.

Planck's time interval

In the primordial series, as soon as two consecutive snapshots have a time interval as short as one Planck's time, then uncertainty occurs and the predictability is impossible. The position in space of the next consecutive snapshot becomes unpredictable. This results in a new state where the material entity of the snapshots can be in different topographical locations at the same Planck's time interval and therefore the distance from the position from its previous position is unpredictable following the equations of the uncertainty principle of Heisenberg at the level of quanta mechanics.

When the interval between two or more snapshots in the primordial monoclonal series occurs within a Planck's time then uncertainty is valid and the governance interchanges to randomicity. The uncertainty occurs because it is impossible to reach to a smaller interval, therefore the synchronous daughter snapshots are happening at the exactly the same time and therefore unpredictability occurs. This means that they become independent from each other notwithstanding that they share a common previous snapshot. Each one of the synchronous daughter snapshots creates a daughter monoclonal snapshot that will evolve independently in time and space.

Daughter monoclonal series

From this point on the primordial monoclonal sequence is split and new monoclonal series of snapshots are generated that we define as daughter monoclonal series. From this interchange point onwards, annihilation of determinism is valid and the governance is instantly random. The synchronous snapshots with the slot interval between them of a single Planck's time will generate new independent monoclonal sequential series of snapshots originating now from daughter monoclonal series that might be located at distance

Snapshots in space

In the case of daughter monoclonal series, two or more snapshots can occur in close to each other locations consecutively in time, as they can also occur distantly. The continuity of the snapshots takes into consideration the time and even if the snapshots occur at close locations, even as close as one Planck's length, or at astronomically distant points, still they belong to the same snapshot sequence series.

Snapshots within the space-time

The first primordial monoclonal snapshot coincides with the start of the Universe at its initial creation. At early time in the evolution (10^{-43} seconds), the actions and the consequent events are occurring extremely rapidly[31] and are compatible with the time clock of snapshots (Planck's time).

---

[31] The presentation of the initial events and snapshots is in the Chapter General Theory of Evolution is in the last Part of this book

The primordial monoclonal snapshot series from the start of the Universe and as long as the interval between the sequential snapshots is more than a Planck's time, the governance is deterministic and predictability is valid. In summary, the governance is deterministic as long as the monoclonal snapshots form a consecutive series and the time interval is more than one Planck's time. As soon as the time interval is equal to a Planck's time, the governance is randomicity in both the primordial and the daughter monoclonal series. Another state when randomicity is the governance is when two or more snapshots occur at different location but at the same Planck's interval.

Polyclonal snapshot series

We define as polyclonal snapshots when from a daughter snapshot series a snapshot occurs at the same Planck time with another snapshot from another daughter series. The result in this case is that again, there is interruption of determinism by randomicity and the two synchronous snapshots belonging to separate and independent daughter series create snapshots that are not anymore sequential: During the time of the interchange in Planck's time, there is an instantaneous switch of the running single monoclonal series to a polyclonal consisting of new series that thereafter each will continue as monoclonal again.

It is not a prerequisite that at each Planck's time a snapshot is registered; there might be intervals with no-snapshot occurring. Therefore, the number of snapshots does not coincide with the number of Planck's times: in the case of the primordial monoclonal snapshots, their number might be less than the number Planck's time. On the contrary, in the case of polyclonal snapshots their number can be higher.

Two or more snapshots can occur in close to each other locations consecutively in time, as they can also occur distantly. The continuity of the snapshots takes into consideration the time and even if the snapshots occur at close locations even as close as one Planck's length, or at distant ones, still they belong to the same monoclonal or polyclonal snapshot sequence.

Complex snapshots

The primordial, daughter and polyclonal series of snapshots involve minute entities such as elementary particles. However, a complex snapshot can regroup larger number of grouped snapshots, such as chemical molecules, objects, or even living organisms or celestial bodies.

In such cases, the observer focuses on the complex snapshot and follows its evolution in time as long as it remains as one grouped entity.

In the case of complex snapshots, the interplay of the five variables influences the loss of determinism during the interchange. At the same time due to the decrease of the importance of quanta mechanics unpredictability the macroscopic world. As the Universe is expanding and the entities increasing multiplicity, complexity and diversity and adjusting accordingly the plasticity and adaptability the distance between the locations of each event to its next sequentially in time event tends to become in most cases distant and even astronomically faraway. Further, more the logarithmically increasing number and types of events consequently shortens the time intervals between the snapshots. The primordial monoclonal series at the first instances after the Big Bang favored series of monoclonal and later polyclonal snapshots.

Snapshot registry

Therefore, in the Universe today, we dispose of a theoretical time registry of all the snapshots since the first instance of the primordial, the daughter monoclonal and the polyclonal series that relate to small entities and in particular the elementary particles. This registry is immeasurably huge and constitutes the natural history of the Universe. The governance is deterministic as long the snapshot series are sequential and interchanges to randomicity during the daughter monoclonal and polyclonal series. Therefore, the evolution maintains a freedom to operate randomly while there is a preservation of determinism as background governance. Enablement of randomicity occurs in addition in all instances in quanta mechanics where the uncertainty principle of Heisenberg is valid. On the contrary, as long as the human mind is absent, there is no indeterministic governance.

Therefore, the evolution at the start of the Universe is deterministic and interchanging at time intervals of the order of Planck's time spontaneously and instantly to random. This continuous switch in governance enables freedom in the process rendering evolution unpredictable, while maintaining predictability by determinism in all snapshots occurring at time intervals longer than Planck's time. Randomicity is at all times valid whenever quanta mechanics are valid due to the uncertainty principle of Heisenberg pertaining in the microscopic quanta mechanics environment.

In the complex snapshot series, the observer selects an object that is made of molecules or larger pieces of objects and follows it in its time evolution as one single entity constituting a special registry of the evolving snapshots. The governance is such a case is deterministic as long as the observer does not engage active Will, then becoming indeterministic. The governance of randomicity in complex series is through the interplay of the five variables.

An example is the observation of the cinematographic snapshot of photographs of the moon as one single entity every one second. The film produced is a complex monoclonal series related to a limited time and a concrete object. In this case deterministic governance is valid. The same is valid if the interval between the photographs is equal to Planck's time. However, on the quanta mechanics of the particles that constitutes the moon object; determinism and randomicity interchange in the monoclonal series involved. The interchanges however, are neutralized between them and in the eyes of the observer the moon is viewed cinematographically as a single object in a time series of complex snapshots. In concurrence in the long-term observable period of the moon its state is stable and therefore the interplay of the five variables is negligible when the moon is examined a single complex object.

Process Philosophy and snapshots

In the evolutionary process described earlier an entity depicted by a snapshot can change by an action which is the consequence of a force and which results to a new snapshot. However, each new snapshot is static at the instance (Planck time) when the event occurs.

In the history of Philosophy Heraclitus statement[32] of "ever-newer waters flow on those who step into the same rivers" describes the consequence of snapshots within the space-time variable. The actions as a result of forces are described already by Empedocles: Love being the uniting force that attracts all entities, thereby creating something new, and Discord as the dividing force that separates and destroys them.

In this sense the proposed evolutionary process in this book is both dynamic following the two pre-Socratics but also static as Plato and Aristotle posited true reality as eternal differing from the Process Philosophy. Process Philosophy identifies metaphysical reality with change and development and regards change as the cornerstone "Being" thought as "Becoming".

---

[32] "ποταμοῖσι τοῖσιν αὐτοῖσιν ἐμβαίνουσιν, ἕτερα καὶ ἕτερα ὕδατα ἐπιρρεῖ"

With one exception, all actual entities for Whitehead are temporal and are occasions of experience. On the contrary we here postulate that all snapshots are "atemporal", unique and self-descriptive.

An additional difference is that the snapshots are not considered as "occasions of experience" related to the subjective (or in the case of God objective) interpretation but are provable events that are reproducible. Furthermore, all the actions, events and resulting snapshots are defined in this book as being under the governance of either determinism, or indeterminism or randomness, on the contrary of the Process Philosophy that considers that such process are never deterministic. We believe that actions are powered depending on the states by either active or passive Will[33].

---

[33] See next Part 2 chapter Will

PART 2 GENESIS OF UNIVERSE

There is no experimental or other evidence in cosmogony[34] to confirm any form of material preexistence prior to the Big Bang[35]. In this context, we will discuss in the next paragraphs, the events that occurred at the early stage just after 10^{-43} seconds from the start, at the quantum era, when the Universe consisted of a soup of leptons and quarks. We will further discuss in a later chapter[36], the earlier than 10^{-43} seconds' events using a novel proposal linked to the materialization of a virtual to a real elementary particle, its implosion and the creation of an initial primary scalar field that triggered the creational automatic process of the Universe.

The Plates 1 and 2 present fields where the value depicted has a space distribution and a time correlation. These Plates show the events after the 10^{-34} seconds up to the appearance of human mind on Earth.

COSMOGENESIS

At the Big Bang, starting from a minimal material energy entity of a minute scalar field[37] under negative pressure an expansive inflation occurred that caused the considerable increase of the mass of the Universe[38] (of more than 10^{53} g) that we observe today. The initial scalar field, that had velocity as energy that activated this minute mass, provoked the start of the inflation.

---

[34] Cosmogenesis is the term used Pierre Teilhard de Chardin to describe the cosmological process of the creation of the Universe. Other processes included biogenesis and noogenesis, culminating to denote the state of the maximum organized complexity (complexity combined with centricity), towards which the Universe is evolving.

[35] This is the event, which led to the formation of the Universe, according to the prevailing cosmological theory of the Universe's early development. The Universe, originally in an extremely hot and dense state that expanded rapidly, has since cooled by expanding to the present diluted state, and continues to expand today.

[36] See Part 4 GENERAL THEORY OF EVOLUTION

[37] The characterization of a field is by a function of position and time with a value at each point.

[38] All matter and energy, including the living organisms, the galaxies, and the contents of intergalactic space, regarded as an entity.

The inflation is the exponential and explosive expansion of the Universe driven by a negative-pressure vacuum energy[39] that lasted from 10^{-36} seconds after the Big Bang to 10^{-33} seconds. The false vacuum[40] is a state of temporary lowest possible energy density and is compatible with any theory that contains scalar fields, that is, fields that resemble electric or magnetic ones, except that they have no direction.

The dimensions of the space of the Universe are increasing exponentially except at the first instance of creation when as postulated in this book, an implosion possibly occurred[41].

The false vacuum disposes of a large negative pressure[42]. In the inflationary theory, the Universe begins as small as 10^{-24} centimeters; which means a hundred billion times smaller than a proton.[43] The inflation takes place while the false vacuum maintains a nearly constant energy density that means that the entire energy increases by the cube of the linear expansion factor, or at least by a factor of 10^{75}.

---

[39] Vacuum energy is an underlying background energy that exists in space even when devoid of matter. The vacuum energy is deduced from the concept of virtual particles, are themselves derived from the energy-time uncertainty principle.

[40] During inflation, while the energy of matter increases by a factor of 10^{75} or more, the energy of the gravitational field becomes more and more negative to compensate. The total energy - matter plus gravitational - remains constant and very small, and could even be exactly zero. Conservation of energy places no limit on how much the Universe can inflate, as there is no limit to the amount of negative energy that can be stored in the gravitational field.
http://web.mit.edu/physics/news/physicsatmit/physicsatmit_02_cosmology.pdf

[41] See Part 4 GENERAL THEORY OF EVOLUTION

[42] Mechanically such a negative pressure corresponds to a suction, which does not sound like something that would drive the Universe into a period of rapid expansion. The mechanical effects of pressure, however, depend on pressure differences, so they are unimportant if the pressure is reasonably uniform. According to general relativity, a gravitational effect is very important under these circumstances. Pressures, like energy densities, create gravitational fields, and in particular, a positive pressure creates an attractive gravitational field. The negative pressure of the false vacuum, therefore, creates a repulsive gravitational field, which is the driving force behind inflation. Mechanically such a negative pressure corresponds to a suction, which does not sound like something that would drive the Universe into a period of rapid expansion. The mechanical effects of pressure, however, depend on pressure differences, so they are unimportant if the pressure is reasonably uniform. http://nedwww.ipac.caltech.edu/level5/Guth/Guth3.html.

[43] The diameter of a proton is estimated to be 1×10^{-13} cm. Illikan, Robert Andrews. Electronics (+ and -) Protons, Photons, Neutrons, and Cosmic Rays. London: Cambridge University Press, 1990: 47

During inflation, while the energy of matter increases by a factor of 10^{75} or more, the energy of the gravitational field becomes more and more negative to compensate[44]. The total energy-matter plus gravitational - remains constant and very small, and could even be exactly zero. The conservation of energy places no limit on how much the Universe can inflate since there is no limit to the amount of negative energy that can be stored in the gravitational field.

This borrowing of energy from the gravitational field gives the inflationary paradigm an entirely different perspective from the classical Big Bang theory, in which all the particles in the Universe (or at least their precursors) are in place from the start. Inflation provides a mechanism by which the entire Universe can develop from just a few ounces of primordial matter[45].

Cosmic inflation

The observable Universe today has expanded to more than 9.3×10^{10} light-years or some 20^{23} kilometers having an average density of 1.0×10^{-31} g per cubic centimeter (or three atoms per cubic meter). The estimate of its atom content of about 9.4×10^{79} atoms gives an approximation for its total mass of the order of 10^{50} to 10^{60} kilograms.

Alan Harvey[46] and Guth were the first to describe the cosmic inflation and stated that the Universe originated in a false vacuum filled with high energy:

"The existence of a repulsive gravitational field caused the Universe to enter a period of expansion. The inflation was so quick that in only 10^{-33} seconds, it was 10^{50} times the original size. Because the false vacuum is not stable, the inflation will not continue forever. Instead, quantum tunneling will cause the false vacuum to decay into a low-energy true vacuum. When it decays, bubbles suddenly seem to fill in the space. Although the bubble Universes start small at first, many of them will

---

[44] http://www.phys.cwru.edu/events/grav_ws/guth-st-thomas.pdf

[45] Published in "The Beamline" 27, 14 (1997).

[46] Alan Harvey is an American theoretical physicist and cosmologist. Guth has advanced the research on elementary particles theories. Currently serving as Victor Weisskopf Professor of Physics at the Massachusetts Institute of Technology, he is the originator of the inflationary Universe theory.
http://web.mit.edu/physics/people/faculty/guth_alan.html

quickly become large. On the contrary to popular belief, the conclusion is that the Universe can suddenly emerge out of nothing."

Guth believes that the Universe exists among countless other Universes with concrete but undisclosed laws of Science. A fractal pattern exists in the multiverse[47] system that involves Universes inside vacuums that are inside other Universes. Each pocket Universe created by inflation will seem flat to the observers who are in it. Meanwhile, new Universes will fill in the gaps created by older ones, similar to Hoyle's discredited steady state theory[48]. The formation of new Universes from the Big Bang is continuous and looks like the cell division in biology without interruption. However, inflation effaces the circumstances of the beginning of the Universe. The Universe according to Guth has a beginning and each Universe is just one of many that emerged. Inflation never ends, but keeps expanding exponentially rate, meaning that it doubles in short increments much less than one second as bubbles in the inflation process.

FOCUS ON FOUR ELEMENTARY PARTICLES

Plate 1 shows the cooling of the Universe in parallel to the inflation starting at approximately 10^{-37} seconds after the theoretical start; however, the first instances of the Universe are yet speculative. This phase transition caused a cosmic inflation, during which the Universe grew exponentially. The current concept is that the Universe is homogeneous and isotropic, which from an extremely high energy density, temperature and pressure, is now expanding and cooling down rapidly.

After its initial explosion and the drop of temperature, the Universe allowed energy conversion to various subatomic particles. Between 3 to 20 minutes after the Big Bang, nucleosynthesis took place followed by matter domination to 70.000 years later. It would take more years for some of these particles to form atoms. Some 370.000 years after the Big Bang hydrogen and helium begin to form as the density of the Universe falls. This occurred about 377,000 years after the Big Bang and after which protons, neutrons, and electrons combine to form other atoms

---

[47] The multiverse is a theoretical framework in modern cosmology (and high energy physics) which presents the idea that there exist a vast array of potential Universes which are actually manifest in some way.

[48] http://physicsworld.com/cws/article/news/2615

(helium and lithium). At the same time, clouds of hydrogen conjugate through gravity to form stars.

In Plate 5 are depicted the events from the early Planck's time to the present. The background is taken from the NASA WMAP Science Team to which are superimposed the four evolutionary tetractides that lead to the eukaryotic cellular DNA and eventually to the human mind.

The gluons are present since the quark epoch of the Universe evolution (10^{-12} to 10^{-6} seconds after the Big Bang), all through the hadrons' epoch (10^{-6} to 1 second) up to the nucleosynthesis. The presence of protons and neutrons occurs during the hadrons' epoch and later during the lepton epoch photons, electrons and neutrinos appear[49].

Here, we focus on four elementary particles amongst all the known fermions and bosons since the Big Bang: these are the quarks u and d, the electrons and the neutrinos. These four particles form the pivotal, tetractide, radical material base in the evolution of the Universe towards a pathway leading to the mind following an anthropic approach[50]. This first distinct tetractide of u and d quarks, electrons and neutrinos, evolves among all other elementary particles during the first minutes of the Universe. Later it is distinctive among millions, or billions or trillions and more of molecules and material forms. Three other distinctive tetractides to the appearance of the human mind are involved later during the evolution to the mind as shown in Plate 2. This Plate focuses on the significance of the tetractides of elements in the evolution of the Universe when focusing on the pathway followed until the appearance of the mind of Humans in the evolution. The same is in Plate 5, where the four tetractides, from the start of the Universe to the human mind, are chronologically depicted.

Neutrinos, notwithstanding that they are not part of the structure of the materials, still they exist abundantly in the Universe, transverse all atoms, from the first instances after the Big Bang until now.

---

[49] http://www.knowledgetreeproject.org/birthoftheuniverse.htm

[50] "To many physicists, the landscape picture is a welcome opportunity to invoke anthropic reasoning: in the multiverse, life will evolve only in very rare regions where the local laws of physics just happen to have the properties needed for life, giving the illusion of intelligent design" Andrei Linde definition.

Focusing on the properties of the Universe that allowed the appearance of life on Earth, we reach to a minimum of sine qua non fundamental physical constants absolutely and essentially necessary to be present.

Plate 3 presents the evolution of the Universe in two levels, up depicts the evolution in the first instances of the Universe and down shows four distinct tetractides of fermion/boson on each line as well as the three generations of fermions. During the time-lapse between 10^{-44} seconds to 10^2 seconds, this interchange of three generations of tetractides correlated with the drop of the high temperatures.

ORIGINS OF LIFE

Life is part of the evolution in the Universe and is a proven situation as observed on Earth. As part of the evolutionary process it is ruled by the laws of Science and governed by determinism, indeterminism or randomness pending on the field of examination. All theological or mystical believes are speculative and outside the spectrum of scientific rational and experimentally proven method.

Life is synonymous with the existence of cellular organisms as provably observed on Earth as part of the Universe at today's time of $13.8x10^9$ years after the Big Bang. The last universal common ancestor (LUCA)[51] is the most recent organism from which all organisms now living on Earth descend from. By focusing the evolution to life within the large process accomplished during this period since the Big Bang the time when Earth finished its transition to being very much like it is today, can be roughly dated to about $4.0x10^9$ years ago (or $9.8x10^9$ years after the Big Bang). Stromatolites, that play a pivotal role in the process of life appearance on Earth, date more than $3.5x10^9$ years ago. The last universal common ancestor is estimated to have lived some 3.5 to $3.8x10^9$ years ago. The concurrent conditions that appear favorable to the compilation of all essential elements to organized biological living organisms must have started at the end of the Hadean geological eon through the Archean eon (4.0 to $2.5x10^9$ years ago) and till the end of the Paleoproterozoic era (2.5 to $1.6x10^9$ years ago) meaning a time span of more than $2.4x10^9$ years.

---

[51] Glansdorff N, Xu Y, Labedan B. The origin of life and the last universal common ancestor: do we need a change of perspective? Res Microbiol. 2009 Sep;160(7):522-8. doi: 10.1016/j.resmic.2009.05.003. The last universal common ancestor (LUCA) is the most recent organism from which all organisms now living on Earth descend.. Nature May 2010) 465 (7295): 219–222. "A formal test of the theory of universal common ancestry" Theobald, Douglas L. Thus it is the most recent common ancestor of all current life on Earth: " A universal common ancestor is at least 10^{2860} times more probable than having multiple ancestors. This is shown in a cladogram of modern taxonomic groups from their common ancestor. http://en.wikipedia.org/wiki/File:Tree_of_life_SVG.svg

During this long period the evolution from essential chemicals under interchanging environmental conditions develop towards the primal RNA of the LUCA through prebiotic chemistry, community and promiscuity that ends to the variable elaboration of a genetic code, of lipids, and of a early form a prokaryote[52] with introns and exon shuff RNA of LUCA. The endpoint of the LUCA arises after infinitely immeasurable innumerous chemical interactions that ended to the uninterrupted evolutionary events as depicted in Plate 2. The four atoms hydrogen, oxygen, nitrogen and carbon were at the core of this interplay that lead to the four RNA nucleotide bases (guanine, cytosine, adenine, uracil) that will evolve to the genetic code of LUCA and eventually to the Humans. The interplay will involve already at the early stage deoxyribose and lipids for the formation of the protocell. Early research[53] raises the fundamentals of this chemical approach.

The search to clarify in the natural history the precise intermediate steps between the early formations of the biosphere on Earth to the appearance of the universal common ancestry and more precisely the LUCA should answer the question "which is the exact pathway followed in the evolution that initiated the abiogenesis to which the Humans and their mind are the most advanced observable endpoint at this time". We propose in this context three options:

- The first is based on the fact that all the necessary building chemical atoms, molecules physical conditions required for the automatic constitution of a protocel (an organism structure, enclosed by a

---

[52] Prokaryotes are organisms that lack nucleus, mitochondria, or any other membrane-bound organelles. In other words, all their intracellular water-soluble components (proteins, RNA or DNA and metabolites) are located together in the same area enclosed by cell membrane.

[53] Szostak et al53 envisaged the creation of synthetic life in the laboratory that simulates the events that occurred during the abiogenesis period of the development on Earth. Recently experimental evidence53 provides positive results in which the properties of amino acids and RNA impose naturally a partially coded polymerization along RNA templates and that associated coding mechanism is remarkably robust against mismatches. When supplied with "meaningful" RNA sequences, translation systems of this kind should be capable of generating pools of proteins a small fraction, which will be functional. The feedback action of these proteins on the translation itself further increases its efficiency, allowing the addition of more codons to its repertoire. Further, the coding regime can naturally occur under prebiotic conditions generating partially coded proteins through a mechanism, which is remarkably robust. The results show that features of the genetic code support the existence of an early translation system and conclude that the natural implementation of abiogenesis model is by a system of four codons and four amino acids, thought to be a plausible original genetic code.

membrane, that carries out some life activities, such as growth and division) up to the late universal common ancestry[54] (LUCA) existed for an extremely long period in the deep-sea hydrothermal vents and in the stromatolites. The evolution managed through numerous chemical pathways to synthesize autoreplicable nucleic codons and protein matrices that lead to the LUCA. It took a long more than $2.4x10^9$ year period of intensive chemical reactions to attain the first living genome able to replicate and have metabolic functions. Specific environments on Earth at the prebiotic era were the deep ocean stromatolites where all the essential atoms and molecules to enable life in an early genome in the existed in the immediate environment abundantly.

- The second is an active interventionism by an external force that ended with the formation of the universal common ancestry. Mythological and religious aphorisms oversimplify the scientific search by intentional obscurantism.

- The third is the galactic option that is only repeating the query from the Earth to another planet or star displacing the query.

The second and third options enable a vicious cycle that does not answer the question but simply relays it to the "which came first - the chicken or the egg?" Other sterile side questions are "who has then created the external intervening action" or in the case of astrobiology "what enabled abiogenesis in the galactic exoplanet" and so on.

In a recent article related to the above questions, Addy Pross and Robert Pascal[55] state that "it seems probably that we will never know the precise *historic* path by which life on the Earth emerged, but, very much in the Darwinian tradition, it seems we can now specify the essence of the

---

[54] The LUCA (last universal common ancestor) is the most recent organism from which all organisms now living on Earth descend. Thus it is the most recent common ancestor of all current life on Earth that is estimated to have lived some 3.5 to 3.8 billion years ago. Nature May 2010) 465 (7295): 219–222. "A formal test of the theory of universal common ancestry" Theobald, Douglas L. The last universal ancestor, also called the last universal common ancestor, is the most recent organism from which all organisms now living on Earth descend. Thus it is the most recent common ancestor of all current life on Earth: " A universal common ancestor is at least 102860 times more probable than having multiple ancestors. This is shown in a cladogram of modern taxonomic groups from their common ancestor. http://en.wikipedia.org/wiki/File:Tree_of_life_SVG.svg

[55] Pross A, Pascal R. The origin of life: what we know, what we can know and what we will never know. Open Biol. 2013 Mar 6;3(3):120190. doi: 10.1098/rsob.120190.

ahistoric principles by which that process came about. Just as Darwin, in the very simplest of terms, pointed out how natural selection enabled simple life to evolve into complex life, so the recently proposed general theory of evolution points out in simplest terms how simple, but fragile, replicating systems could have complexified into the intricate chemical systems of life. But, as discussed earlier, a detailed understanding of that process will have to wait until ongoing studies in systems chemistry reveal both the classes of chemical materials and the kinds of chemical pathways that simple replicating systems are able to follow in their drive towards greater complexity and replicative stability".

We in this book believe that the exact process pathway will be experimentally reproduced as soon as the exact LUCA genome is identified and this proof is underway.

Pathway leading to the LUCA

Probably during the period of more than 2.4×10^9 years of compatible evolutional history on Earth, the first living organisms emerged in deep-sea hydrothermal vents and in stromatolites[56]. This is achieved primarily by the hydrogen-saturated alkaline water meeting acidic oceanic water at underwater vents that would produce a natural proton gradient across thin mineral "walls" in rocks that are rich in catalytic iron–sulphur minerals. This set-up could create the right conditions for converting carbon dioxide and hydrogen into organic carbon-containing molecules, which

---

[56] The definition of stromatolites is that they are laminated accretion structures with a synoptic relief formed in shallow water. The realization of this process is by the formation of biofilms of microorganisms, especially cyanobacteria that trap, bind and cement sedimentary grains. Stromatolites-building communities include the oldest known fossils, dating back some 3.5×10^9 years when the environments on Earth were too hostile to support life, as we know it today. The age of the oldest fossil microbe-like objects is 3.5×10^9 years back (or 10.2×10^9 years since the Big Bang) and approximately 1.0×10^9 years after the formation of the Earth itself.

The stromalites include the most ancient records of life on Earth and constitute the ideal substrate milieu for the fermentative promiscuous prebiotic soup reactions. Prokaryotic bacteria, including cyanobacteria and Archaea, form an important part of the biomass thriving in extreme environments owing to their relative lack of complexity.

Recent experimental evidence advances our knowledge towards the resolution of the process of the emergence of life on Earth in the mid of unfavorable environmental and resource conditions. More precisely Paul Clarke[56] et al showed that precursors of amino acids, the building elements of proteins, catalyze the formation of the skeleton molecules of DNA and RNA under prebiotic conditions amid the stromalites.

can then react with each other to form the building blocks of life such as nucleotides and amino acids.

Secondly, it can be achieved by the protocell membrane that defines a spatially localized compartment and an informational polymer that allows for the replication and inheritance of functional information. Recent studies of vesicles composed of fatty-acid membranes have shed considerable light on pathways for protocells growth and division, as well as means by which protocells could take up nutrients from their environment. Additional work with genetic polymers has provided insight into the potential for chemical genome replication and compatibility with membrane encapsulation.

Protocell, RNA replication, vesicle growth and division

Simple physicochemical properties of elementary protocells can give rise to essential cellular behaviors, including primitive forms of Darwinian competition and energy storage. Such preexisting, cooperative interactions between the membrane and encapsulated contents could greatly simplify the transition from replicating molecules to true cells.

The theoretical protocell is made up of only two molecular components, a RNA replicase and a fatty acid membrane. An extremely pared down and simple version of a cell, the protocell is nonetheless capable of growth, replication, and evolution. Although a working version of a protocell has not yet been achieved in a laboratory setting, the goal appears well within reach.

The protocell includes two or more RNA replicases which are able to make copies of each other. Concurrent with RNA replication, the vesicle membrane grows through the addition of fatty acids from micelle collisions. This causes the surface area of the protocell to increase while the volume remains constant, resulting in the elongation and increased instability of the protocell membrane. The membrane eventually divides, forming two daughter protocells, with the RNA replicases randomly divided between them.

Every once in a while, a replicase will make a mistake, and a mutant replicase RNA is produced. Usually, this mutation will result in a poorer replicase, if catalytic activity is retained at all. Rarely, however, a better replicase could be formed – a replicase that might be able to copy RNAs faster, for example. Even more rarely, RNA will be introduced to a protocell (by mutations or some other means) that has a new, different catalytic activity, such as the ability to catalyze the formation of fatty acids. Recently[57,58] research in this direction support the pathway leading from abiogenesis to the LUCA and finally to the Humans, has been published. The protocells with faster replicases or new functional ribosome will have an advantage over other protocells by being able to grow and divide faster.

Randomicity versus determinism

We can provisionally conclude at this point based on the present knowledge, that the first living organisms are probably unicellular DNA eukarya that emerged by endosymbiosis in parallel to unicellular archaea and bacteria that emerged by reduction[59]. All these living cells derived

[57] Powner MW, Gerland B, Sutherland JD. Synthesis of activated pyrimidine ribonucleotides in prebiotically plausible conditions. Nature. 2009 May 14;459(7244):239-42. doi: 10.1038/nature08013.

"At some stage in the origin of life, an informational polymer must have arisen by purely chemical means. According to one version of the RNA world hypothesis this polymer was RNA, but attempts to provide experimental support for this have failed. In particular, although there has been some success demonstrating that 'activated' ribonucleotides can polymerize to form RNA; it is far from obvious how such ribonucleotides could have formed from their constituent parts (ribose and nucleobases). Ribose is difficult to form selectively, and the addition of nucleobases to ribose is inefficient in the case of purines and does not occur at all in the case of the canonical pyrimidines. Here we show that activated pyrimidine ribonucleotides can be formed in a short sequence that bypasses free ribose and the nucleobases, and instead proceeds through arabinose amino-oxazoline and anhydronucleoside intermediates. The starting materials for the synthesis—cyanamide, cyanoacetylene, glycolaldehyde, glyceraldehyde and inorganic phosphate—are plausible prebiotic feedstock molecules, and the conditions of the synthesis are consistent with potential early-Earth geochemical models. Although inorganic phosphate is only incorporated into the nucleotides at a late stage of the sequence, its presence from the start is essential as it controls three reactions in the earlier stages by acting as a general acid/base catalyst, a nucleophilic catalyst, a pH buffer and a chemical buffer. For prebiotic reaction sequences, our results highlight the importance of working with mixed chemical systems in which reactants for a particular reaction step can also control other steps"

[58] Bianconi G, Zhao K, Chen IA, Nowak MA. Selection for replicases in protocells. PLoS Comput Biol. 2013 May;9(5):e1003051. doi: 10.1371/journal.pcbi.1003051. http://www.sciencemag.org/content/314/5805/1558.full?maxtoshow=&HITS=10&hits=10&RESULTFORMAT=&fulltext=irene+chen&searchid=1&FIRSTINDEX=0&resourcetype=HWCIT

[59] http://www.biology-direct.com/content/3/1/29. Glansdorff N, Xu Y, Labedan B. The last universal common ancestor: emergence, constitution and genetic legacy of an elusive

from organic molecules surrounded by a membrane-like protocell structure.

The observations in the above paragraphs show that life originated and emerged from cell membrane bioenergetics in chemistry of deep-sea hydrothermal vents. The promiscuity in the primordial prebiotic soup in the stromatolites constitutes another factor that lead to development and differentiation of the early forms of life.

From the LUCA starts the Darwinian evolution that lead to the apparition of Humans on Earth. The achievement of the endpoint, the mind and the human societies, relied on a probabilistic approach based on randomness and determinism.

The time taken allows the development of enormous numbers of chemical combinations; every developmental stage in multiple space-times in a close promiscuity increased the combination potential of all five variables (multiplicity, complexity, selectivity, adaptability, plasticity). This in turn increases the uncertainty within the randomicity governance reducing determinism and predictability and giving freedom to selectivity, adaptability and plasticity.

It is impossible to define the exact instance of the creation of the LUCA. Focusing on the level of the organic molecules that led to replicable nucleotides the estimate of the span of time when the biochemical reactions occurred is of the order of more than 2.4×10^9 (and the closest estimate of the appearance on earth of the LUCA is 3.4×10^9 years ago). The Shapiro[60] primordial soup is included in this time span.

This long period allowed infinite number of biochemical interactions and the governance was deterministic and random on the particle; it was deterministic taking also into account the five variables on the atomic and molecular level. This resulted in incalculable polyclonal and composite

forerunner. Biol Direct. 2008 Jul 9;3:29. doi: 10.1186/1745-6150-3-29. Eukarya: Eukarya (or Eukaryota) is one in the three-domain system of biological classification introduced by Carl Woese in 1990. The other two are Archaea and Bacteria.
[60] The early Earth had a chemically reducing atmosphere. This atmosphere, exposed to energy in various forms, produced simple organic compounds. These compounds accumulated in a "soup", which may have been concentrated at various locations (shorelines, oceanic vents etc.). By further transformation, more complex organic polymers – and ultimately life – developed in the soup. Shapiro, Robert (1987). Origins: A Skeptic's Guide to the Creation of Life on Earth. Bantam Books. p. 110.

snapshots that gave beyond measure synthesis of different molecules in immeasurably changing reaction conditions. High[61] promiscuity conditions in stromalites favored monoclonal and polyclonal snapshots occurrence at extremely short time intervals escalating the randomicity and allowing at one or more points the emergence of nucleic acid automatic replication process. This finally, led to the emergence of the first living organisms and eventually to the LUCA on Earth.

Determinism and randomness role

If we compare to the plasma-gluon soup of matter that existed only split-seconds after the Big Bang and the prebiotic primordial soup the main difference is the length of the time span lasted: the first extremely short on the contrary the second lasted more than 2.4×10^9 years. The first exists at extremely high temperature and/or density with free quarks and gluons the second at 37 to 70° Celsius and includes myriads of molecule combinations of ninety eight chemical elements. The first soup is dominated by randomness because of the uncertainty of Heisenberg's principle the second by the interplay of the five variables.

The concurrent command of deterministic and random governance based on the relation uncertainty, provably ended to the living organisms that appeared during the evolution up to now. This meaning that there is no certainty if in exactly similar conditions the endpoint of reaching to a living organism will be again reproducible elsewhere, as it happened on Earth or that on the contrary that it has occurred one or more times on other exoplanets. It also means that the evolution because of the causative uncertainties is not exclusively deterministic and depends on randomicity. The randomicity is insignificantly influenced in the macrocosms by the quanta mechanics environment due to the short intervals between snapshots which renders dominant the interplay of the five variables in the molecular and macroscopic world.

---

[61] Eleven taxa (including eight heretofore undescribed species) of cellularly preserved filamentous microbes, among the oldest fossils known, have been discovered in a bedded chert unit of the Early Archean Apex Basalt of northwestern Western Australia. This prokaryotic assemblage establishes that trichomic cyanobacterium-like microorganisms were extant and morphologically diverse at least as early as 3.4×10^9 years ago and suggests that oxygen-producing photoautotrophy may have already evolved by this early stage in biotic history. Schopf JW. Microfossils of the Early Archean Apex chert: new evidence of the antiquity of life. Science. 1993 Apr 30;260:640-6.

EVOLUTION TO HUMANS

Emerging from the LUCA as discussed earlier the evolutionary tree shows the divergence to the contemporary species of bacteria, archaea and eukaryotes. The eukaryotes evolve to protozoa, algae, plants, moulds, fungi and finally animals. Determinately and by Darwinian evolution the process ends to the hominoids and the Humans that develop mind and human contemporary societies.

The laws of Science rule all actions, events and subsequent snapshots of the evolution of the materials including living organisms according to the selection of species and to the biological homeostasis. Since the Big Bang up to Humans, the evolution is compliant to the mentioned series of tetractides at each echelon. The last tetractide (Plate 2) in the chain of the biological evolution is the four-nucleotide interplay in the genetic code of all living organisms.

Humans are, in the observable Universe, the most advanced entity that evolved following an uninterrupted chain of genetic mutations that occur under the governance of determinism and randomicity principally reigned by the interplay of the five variables. In particular, on the genetic level and the nucleotide tetractides the interplay of the five variables and the consequential gene mutations play an important role in the Darwinian evolution. The introduction of indeterminism occurs just after the emergence of Humans as the third governance.

The purposes in the evolution of the Universe until the appearance of the mind are passive under the governance of determinism and occasionally random. Randomicity is coincident with uncertainty that is provoked when quanta mechanics are valid or when the interchange of the intervals between snapshots reaches Planck's time. Randomicity is also valid when the interplay of the five variables provokes uncertainty. Darwinian evolution is under the governance of determinism and with the interplay of the five variables of randomness.

Like all living organisms, Humans are evolving based on genetic inheritance and environment as interplayed by the five variables. The evolution in general and especially of the mind is not a one-time snapshot: it consists of a series of evolutionary processes that occur at various occasions in the space and time. In addition, there is an asynchronous and differentiated evolution on Earth since 2.3×10^6 years of the different species and sub-species of the genus *Homo*.

Role of indeterminism

The incursion of the governance of indeterminism is due to the Human mind. During the interventionism by the mind, the active actions (and respective snapshots) are unpredictable and independent of the antecedent one. The evolution with the appearance of the mind creates an era in the Universe that opens the possibility of active interventions that interfere on the deterministic passive actions, events and the consequential snapshots describing each change in the space-time related to any single movement.

The sequential series globally remain deterministic or random, up to the incursion of mind governance indeterminism. As long as indeterminism governance is inactive, randomicity is valid whenever uncertainty occurs. With the increasing diversity of the interplaying events (five variables), the interval between snapshots is reduced and will drop to an interval equal to Planck's time rendering the system unpredictable. The mind powered by active Will has the intervention potential to interrupt at any time the sequential continuity of the snapshots; imposing indeterminism for active purposes and active actions.

The intervention of the mind results in active actions that can modify the deterministic governance. These interventional actions under active Will circumvent both determinism as well as randomness caused by uncertainty relations.

The mind has developed functions, programs, methods and tools that enable active actions powered by the active Will that aim to active purposes in an indeterministic process that annihilates predictability and randomness. In such a case, the mind acts by strategic interventionism. In view for an active intervention action, the mind should be operational, meaning developed, integrated, vigilant, alert, attentive and affirming consciousness and interventionism.

Evolution purposes

The differentiation of active and passive purposes and their potential actions are in relation to the governance. Table 1 presents five examples in the evolution. More precisely, example (1) refers to the start at zero seconds corresponding to the Big Bang[62]. Example (2) refers to the

---

[62]See Part 4 GENERAL THEORY OF EVOLUTION the first causative instance of the Big Bang in relationship to an explosive expansion as the primary scalar field.

formation of particles, electrons captured by nuclei to form atoms; Example (3) refers to the emergence of the LUCA (late universal common ancestry organism). Example (4) refers to the emergence of the primate *Pan Troglodytes* and finally example (5) to the human mind when operational.

TABLE 1 EVOLUTION VARIABLES TIMELINE

TIME	TIME AFTER BIG BANG			YEARS AGO	
	10^{-43} sec	$3x10^4$ sec	5×10^9 years	$6x10^6$	$3.5x10^4$
EXAMPLES/ CONDITIONS	1	2	3	4	5
Active purpose	$\varnothing$	$\varnothing$	$\varnothing$	$\varnothing$	+
Active Will	$\varnothing$	$\varnothing$	$\varnothing$	$\varnothing$	+
Active action	$\varnothing$	$\varnothing$	$\varnothing$	+	+
Passive purpose	+	+	+	+	+
Passive Will	+	+	+	+	+
Passive action *	+	+	+	+	+
Multiplicity	>>>	>>>	>>	>	>
Complexity	>	>>	>>>	>>>>	>>>>
Diversity	>	>>	>>>>	>>>	>>
Plasticity	$\varnothing$	>	>>	>>>	>>>>
Adaptability	>	>>	>>>	>>>>	>>>>
Snapshot intervals	>>>	>>	<<	<<<	<<<<
Dimensions of Universe**	<<<	>	>>	>>>	>>>
Determinism	>>	>>	>>	>>	>>
Randomness ***	>>>>	>>>	>>	>	>
Indeterminism	$\varnothing$	$\varnothing$	$\varnothing$	$\varnothing$	+++

EXAMPLE CONDITIONS:

1. Big Bang, 2. Elementary particles, 3. Late Universal Common Ancestry (LUCA),

4. P. troglodytes, 5. Human mind

Y years since Big Bang, sec means seconds, >> Large   << Small, + Yes, Ø No, ± can occur or not occur pending to the variables

* Effective through the central nervous system that manages voluntary active actions (pyramidal) and involuntary (extrapyramidal) responses in Humans and some animals where applicable as action to external stimuli i.e. chemical, electric, or neuromuscular response

** At the creation, there is an implosion of the first particle followed by the expansive inflation (See Part 4 General Theory of Evolution).

*** Randomness in the early instances after the Big Bang is mainly based on the uncertainty principle of Heisenberg, later in the macrocosms, the interplay of the five variables are more dominant however causing less randomness. The mind when acting using active Will is practically free from randomness since uncertainty due to Heisenberg's principle or the five variables is insignificant.

Passive purposes, passive Will and passive resulting actions are valid since the start of the Universe, under the governance of determinism and ruled by the laws of Science. In living organisms, they are responses to specific stimuli. In more advanced organisms in which communication organs are developed, passive actions result from central nervous system responses. Passive actions are involuntary in respect to the mind; on the contrary, all active actions powered by the active Will are voluntary.

Every achieved action is an event as the praxis of a response and constitutes a concrete snapshot. In biology, the stimuli provoke respective specific responses. Specifically in higher animals where intervention of the central nervous system occurs, the response is either passive (involuntary, automatic) through extra-pyramidal nerve circuits or active (voluntary) through pyramidal. In such a case both passive-involuntary and active-voluntary, responses are under the rule of passive Will. Only the mind circumcises all passive actions and under the power of the active Will proceeds to active actions towards active purposes.

Active purpose, active Will and consequent active actions are only applicable after the mind is operational and in terms of evolution history only after the emergence of Humans.

GOVERNANCE

The five variables' interplay causes uncertainty related to the snapshots resulting from each action. In Table 1 each one of the five variables is adjusted to the timeline of the evolution from the Bing Bang to the appearance of the Humans on Earth with reference to the five examples described as to the variables (Active purpose, active Will, active action, passive purpose, passive Will, passive action, multiplicity, complexity, diversity, plasticity, adaptability, snapshot intervals, dimensions of Universe, determinism randomness and indeterminism). The scoring of these variables is arbitrary however their effect on the evolutionary timeline of the Universe shows their impact on the unpredictability and randomness of the process.

The variables are prominent as the complexity of the evolution advances and in particular as more events from action are occurring concurrently and at different locations. The interplay of the five variables contributes to uncertainty and increases the effect of the governance of randomicity in the macro-cosmos where the laws of Science are applicable. There is a reduced uncertainty at the first instances of creation when the variety of elementary particles is scarce despite their large numbers.

As the multiplicity of the evolution advances in time, there is a decreased interval between snapshots reaching to Planck's time. This causes an increased uncertainty and randomicity due to the creation of polyclonal snapshots on the quanta mechanics level. At the first instances, the intervals are longer; however, the rapidity of the evolving events in space in the first instances minimizes the importance of their time length.

Determinism is present although the evolution of the Universe and after the emergence of the mind is concurrent with indeterminism. The randomness, where indeterminism is concurrent with the emergence of the mind is minimal. As the evolutionary timeline advances randomicity is more dependent to the interplay of the five variables (macro-cosmos) than on the uncertainty of the shift of snapshots (quanta mechanics).

Examples in evolution

Table 1 summarizes the five examples interposed with influencing conditions (active or passive purpose, Will, action, the five variables, and the time among shifting snapshots, the dimensions of space where event is occurring, determinism, randomness and indeterminism).

The example (1) refers to the first instances of the expansive inflation that followed the start. The description of the start refers to the time from creation to the Big Bang or more precisely to 10^{-43} seconds after the start and is described together with the time from the start to 10^{-43} seconds in the last Part of this book[63].

---

[63] See Part 4 General Theory of Evolution

In example (2), the formation of particles during the massive explosion involves movement where the conditions are interplaying on the outcome of each individual action and depends on the snapshot shifting in time-space.

For the example (3) and (4), that is the universal common ancestry and the appearance on Earth of the primate *Pan troglodytes*, the closest relatives to Humans, the laws of Science will be valid on the macroscopic level. The governance will be deterministic globally and only on the elementary particle level, quanta mechanics and randomness will occur. Randomness on the quanta level is insignificant in the macro-cosmos however the interplay of the five variables is significant.

The differentiation of Humans and the mind is one-step further, still being very close to the evolutionary step of monkeys especially as to the central nervous system. Examples (4) and (5) present the differentiation of primates and of Humans from the other animals. The example (5) relates to the human mind that can perform active actions, responding to active purposes. Active Will as the mobilizing force, powers these active actions. Concomitantly Humans share with other living organisms the passive biological purposes, actions and Will[64].

Humans as living organisms in the Universe

Humans as living organisms[65] (*Homo sapiens sapiens*) exist in the biosphere of the Earth and therefore are parts of the Universe and live as individual entities according to the laws of Science (including the Darwinian principles of the evolution of species) functioning and operating with actions. Humans, as all material entities, share pertained purposes, dispose attributes and perform actions. Every movement (action) of each atom of each Human consists of a separate snapshot that is part of a series of other snapshots registered of each other material entity in the Universe.

The evolution, Humans appear already before 1.6 to 1.9×10^5 years ago however, emerging at diverse timings in the continents on Earth. The mind has developed gradually and concurrently with the progress of

---

[64] As described earlier active Will is the force that will succeed to an achieved active action toward an active purpose. On the contrary the passive Will commands a passive action toward achieving a passive purpose and finally a passive action.

[65] Kingdom: *Animalia*, Phylum: *Chordata*, Class: *Mammalia*, Order: *Primates*, Family: *Hominidae*, Subfamily: *Homininae*, Tribe: *Hominini*, Genus: *Homo*, Species: *H. sapiens*, Trinomial name: *Homo sapiens sapiens*

knowledge, science, art as well as ethic and activities compliant and reliant to the society where living.

Purposes of Humans

The mind is, in the observable Universe, the most advanced achievement of the evolution since the Big Bang. Empirically, this achievement, when considered with a human-mind centric approach, is the global purpose of the evolution of the observable Universe now. Specifically for Humans, that dispose already mind, the purpose is the advancement of the level of mind.

Specifically the purposes of Humans and their societies are in the frames of those of the Universe constituting an evolutionary event. The purposes are active or passive: the overall purpose of the evolution is to reach the level of an always-superior human mind; this statement is based on the natural history of evolution from the Big Bang to the Humans on Earth.

Attributes

The description of Humans, as individuals, defines them by their dynamic attributes that develop during the biological life in a given space-time. The attributes depend from the particular individual's biological, anatomic and physiologic infrastructure and from the experiences acquired in the space-time where living. The basis of each individual's infrastructure is the genetic inheritance combined to the environmental surrounding.

The real existence of a Human as an entity in a space-time proves his or her concrete ontological existence. This existence evolves following a continuous series of events that reflect on each of the sequential lifetime snapshot shifting from conception to death.

Modules of Humans

The two modules that form Humans are the mind and the body. Both are material from the start of the individual life through death, living continuously as reflected by each sequential snapshot. The environing material entities confront at each instance Humans as well as human societies in general. The environment of Humans includes the other individuals in the biosphere, celestial bodies, and buildings and manufactures of human societies as well as globally all the elementary particles, energy and radiation existing in the Universe.

Humans dispose a complicated information system with constant communication with their body and mind as well as with the surrounding time-space. In Plate 6 presents the functions of the two modules of Humans as well as the architectural structure of the human mind in relationship to its modules.

The mind functions are the privilege of Humans; both animals and Humans share the body functions and operate through the same anatomical and histological infrastructure, that is the body cells, tissues, and organs anatomical infrastructure.

Potential action and achieved action

The potential actions at each space-time point are multiple and the prevailing decisive achieved action at each moment is under the influence of the environmental conditions, the passive as well as in Humans the active intervention actions of the mind through the command of active and passive Will respectively. Humans through active Will have the choice of the possibility to select the final action, among several potential options at a given space-time. The achieved action can be passive or active.

All actions are functioned through the combined use of the body and the mind in a coordinated operation powered by the active and passive Will and satisfying active and passive purposes.

The performance of an action as a response to an active purpose using active Will through the interventionism of the mind is under indeterministic governance (Plate 7). On the contrary, when the action responds to a passive purpose then it is under deterministic governance. The entire body that includes its cells, tissues and organs operates both the bodily and the mind functions of Humans.

Through passive Will (deterministic), Humans and certain other living organisms (as shown in Plate 7) aim primarily to their survival and reproduction as living organisms. Again, Humans and animals that dispose central nervous system (CNS) as well voluntary movement pyramidal tracts have the ability to proceed to active purposes and active actions that are exceptionally indeterministic, as a response to instincts related to survival and reproduction. However, this animal ability differs from Humans that are dowered with mind and proceed to active purposes and actions aiming by the aesthetic, intellectual and psychological functions to perceive, verify and respond to scientific, aesthetic/artistic, ethics/activities notions.

In Humans, the hierarchy of responses to stimuli can be voluntary and carried with active actions for an active purpose and be indeterministic as shown in Plate 7. In the same Plate are depicted in Humans again, the responses to acquired religious, social, familial, tribal, nationalistic, ethnic, moral, political or fanatical impetuses or taboos aiming to passive purposes and respectively passive actions powered by passive Will and under deterministic governance.

This case raises questions on the legality to judge a person for actions that were performed by such responses. Is for example an act, considered criminal by the collective mind of the society that is executed under the influence of fanatical impetuses "heroic" or reprimand? In this case the "hero" executed the act under the power of passive Will and his purpose and action is passive. The response to a passive purpose is processed with passive actions that are under the governance of determinism.

The pyramidal reflexes are the verge and simultaneously the common field for the integration of responses to stimuli pending on the governance of either indeterminism or determinism from one hand or from active or passive Will from the other. The defensive movement of hand for the defense from a danger can be under passive Will and carried out involuntarily using the same neuromuscular cells as when an active action under active Will is processed. The same act in Humans can be under the active Will when the individual before running sounds the alarm and helps evacuating the space before escaping.

WILL

We propose the Will as the unified force that powers under the governance command, every move or change in the Universe that ends-up by a new sequential snapshot. Humans constitute an integral part of the Universe and therefore powered by the Will.

The unification of the four fundamental forces of the Universe[66] (Grand Unified Theory, GUT) theoretically could have existed very early at the Big Bang start. This is the aim of several hypotheses and running experiments in Physics trying to prove the unified quanta field theory.

The active or the passive Will power the active or the passive action respectively. In both cases of active and passive action in an anthropocentric and mind-centered approach, the empirical purpose is the ascending evolution of the mind.

The particularity of active Will in Humans is that it is governed by indeterminism which is introduced for the first time in the evolution of

---

[66] In physics, a fundamental interaction or fundamental force is a mechanism by which particles interact with each other, which cannot be explained in another interaction.

the Universe; at least as we observe from the standpoint of the Earth. Active Will is mind-related and is as well as indeterministic independent, interventionist, conscious and voluntary. It relies on the level of development of the mind, the education, the level of civilization of human societies and the environment. Humans using their mind functions can master active actions powered by active Will but remain submissive and subservient to passive Will like the rest of the entities in the Universe whenever active Will does not infringe.

Passive and active Will is under the rule of the laws of Science. Passive Will is deterministic, automatic and predictable not requiring external intervening action and is compliant to the passive purposes of the Universe.

Questions as to the role of Will in evolution

We hypothesize in this book that the Will, as observed empirically and anthropocentrically, commands the process and implementation of the ascending trend of the Universe from the Big Bang to the mind. This is shown in Plate 2 where the curve Z represents this hypothesis.

In this Plate the ascending empiric curve Z illustrates the timeline from the Big Bang to the human mind and projects the future (13.8×10^9 years after the Big Bang). Will this ascending trend observed in the Universe since the Big Bang to the human mind continue to ascend? Will the involvement of the active Will of Humans contribute to a potential continuation of the ascension in the future? The answers are speculative since depending to the governance of randomicity. Probabilistically the increasingly calculated number of exoplanets and the plasticity and adaptability of the living organisms combined with favorable physical, cosmological and environmental conditions eventually can contribute to such an ascending Z curve in the future.

Action and Will

Human mind powered by active Will has the liberty to intervene by indeterminism and selectively powering for one or the other of many potential actions. In addition, they dispose reflexes responding to passive purposes, which are involuntary, toward passive actions that are deterministic. In this sense active Will bypasses passive Will and passive actions.

Several animals that dispose a central nervous system (CNS) with the pyramidal tracts able to effectuate voluntary movements can perform voluntarily actions (Plate 7) albeit powered by passive Will. On the contrary, active Will powers the active voluntary purposes and actions exclusively of Humans.

Plate 7 shows that the active Will is only active through the human mind. In contrast passive Will powers all entities in the Universe including also Humans. The reflex responses are common in Humans and some animals and remain deterministic. The reflexes as well as intuitive and instinctive responses of Humans and of some animals are passive and involuntary powered by passive Will.

In animals and Humans that dispose central nervous system, the responses to stimuli that trigger reflexes are involuntary and deterministic through the activation of extrapyramidal[67] neurons. Passive Will powers, in Humans and in some animals', voluntary deterministic actions, achieved through the pyramidal[68] neurons. As discussed earlier, human mind powered by active Will powers active actions that voluntary and under the governance of indeterminism.

Active and passive actions and responses

Both the pyramidal and extrapyramidal neurons are part of the central nervous system that Humans and certain animals dispose, notwithstanding in primates and Humans being more advanced and complicated. (Plates 6 and 7).The responses through the pyramidal system are voluntary and their operation is through the intervention of the central nervous system[69] in bilateral animals. The voluntary response differs between Humans and animals: in animals it is under the power of passive Will while in Humans it is either under passive or active Will.

---

[67] The extrapyramidal system dampens erratic motions, maintains muscle tone and truncal stability. It is phylogenetically older that the pyramidal system and thus plays a more important role in lower animals. It is also part of the motor system involved in the coordination of movement. The pyramidal pathways may directly innervate motor neurons of the spinal cord or brainstem, whereas the extrapyramidal system centers on the modulation and regulation (indirect control) of anterior horn cells.

[68] The pyramidal tract is a massive collection of axons that travel between the cerebral cortex of the brain and the spinal cord. The pyramidal motor system controls all of our voluntary movements. The pyramidal tract mostly contains motor axons.

[69] The central nervous system (CNS) is the part of the nervous system that integrates the information that it receives from, and coordinates the activity of, all parts of the bodies of bilateral animals—that is, all multicellular animals except sponges and radically symmetric animals such as jellyfish. It contains the majority of the nervous system and consists of the brain and the spinal cord, and the retina.

A voluntary response is an active response that is interventional, however the presence or absence of the mind, differentiates the respective responses. For example, Humans and *Pan Troglodytes*, the closest primate to Humans, both dispose the capacity of voluntary pyramidal responses; however, Humans operate the response through both the central nervous system like *P. troglodytes* (passive Will) but in addition through the mind (active Will). Another important point is that both *P. troglodytes* and Humans, in the case of passive Will, the empowerment is under the governance of determinism. Thus, both Humans and *P. troglodytes* will use their central nervous system to achieve the response to stimuli like the escape from a danger or to calculate a strategy of actions to respond to an acquired societal or moral impetus, a taboo, or a physical need in the absence of all functions of the mind. However, Humans through the mind under the governance of active Will, investigate, find the cause of the fire and eventually act differently from escaping the imminent danger. Concurrently Humans and other animalia are processing involuntary passive actions, that are deterministic and that have passive purposes, powered by passive Will.

The intervention of the mind is highlighted with the following example: a philosopher while observing an apple for example will seek the truth behind the desire of simply eating the fruit as would do passively a vegetarian primate or a primitive Human. The philosopher will investigate the reasons that underlie behind the beauty and the appetizing qualities of the apple.

DEFINITIONS

- The mind is an information functional system formed from a cellular biological bodily infrastructure and from information notions. The contemporary mind is extended to external to the body computer hardware and software. The purpose of the human mind is to maintain constant communication of information between the environment and the body and within the body

- The definition of the mind in this book is restricted exclusively to Homo sapiens and only from the instance when it is in full function

- The mind operates using standard functions, programs, methods and tools. The three functions of the mind are the aesthetic, the intellectual and the psychological

- The mind operates with its functions, programs, methods, tools and instruments has as an object the processing of information notions

- The information notions are numerical signals conveyed and expressed through biochemical, physicochemical, electrical or mechanical signals. In the case of mind extension to computer the signals are electronic. The signals in all cases are translatable and express and correspond to material or immaterial notions

- The correct capture of a notion as a signal is achieved through the perception and verification programs that lead to a resolute and concrete response or action through the response program. The notion as a signal indicates, instructs, warns, commands or reduces the level of ignorance of the receiver

- The mind performs its three functions interwoven to the existential individual life of a Human, as a living organism in space-time

- The functions process information between bodily cells, tissues, organs and the environment

- All the functions of the mind are built-up through the information communication between from one hand the cells, the tissues and the organs of the body and from the other the intellectual and psychological and aesthetic functions of the mind and vice versa

- The somatic sensorial functions are referring to the sensory cells, tissues and organs related to the senses and performed by transmitting impulses from sense organs to a nerve center

- The philosophical models are the idealistic and the materialistic as well as the information which is introduced in this book

- Will is the moving force that powers all actions occurring in the Universe. Will is either active or passive

-   All actions in the Universe, either active or passive, have a purpose.

-   Determinism, indeterminism and randomness constitute the governances that command all states in the Universe

- Active Will powers all active purposes that have indeterministic governance, allowing voluntary active actions

- The categories of notions are the true (real) states and scientific, those of aesthetics (art) and those of ethics and of activities compliant to ethics

- The true (real) states notions are those that refer to existing activities or entities

- Scientific notions are those that are prone to experimental proof

- Aesthetics, as well as ethics and activities notions are those that are prone as to their beauty, justice, virtuosity, wisdom or prudence

- Aesthetic notions focused on art include painting, film, photography, sculpture, other visual media, music, theatre, dance, literature, and interactive media and include notions not only on art but also in general of culture, humanities and nature

- Verification is the process of how the mind at various evolutionary stages can authenticate, validate, certify and prove the correctness of each notion

- Since the appearance of the human mind is evolving stepwise trending to higher levels from the individual, to the collective, to the advanced and finally to the superlative

- Notions: are independent, autonomous information signals constituting individual entities that can be either material or immaterial made from elements or systems of elements transmitted with a sign or with a combination of signs as part of a concrete subject matter or archive. The notions are signaling entities either material (electric, chemical, physical, mechanical or other) or abstract immaterial messages

- The aesthetic functions refer to the judgment, interpretation and appreciation of the nature of sensations of both the body and the environment as captured, perceived, felt and processed by the mind define

- The bodily (somatic) sensorial functions respond to aesthetic stimuli perceived by the sensorial organs as reflexes and are distinct from the aesthetic functions of the mind

- The processing of the aesthetic functions of the mind is by evaluating and responding the perceived aesthetic notions[70]. The achievement of this is by empowering active Will and searching for the reason, cause and truth that underlie these notions

- Intellectual functions are related to intelligence as defined in the Main Stream Science of Intelligence[71] as notions related to the ability to reason, plan, solve problems, think abstractly, comprehend complex ideas, learn

---

[70] Following the classical philosophers, aesthetically appealing objects are beautiful in and of themselves. Plato states, "Beautiful objects incorporate proportion, harmony, and unity among their parts". Similarly, in the Metaphysics, Aristotle found that the universal elements of beauty were order, symmetry, and definiteness. (From the definition given of aesthetics in the Merriam-Webster Dictionary Online)

[71] http://www.udel.edu/educ/gottfredson/reprints/1997mainstream.pdf

quickly and from experience in a broader and deeper capability for comprehending the surroundings as well as the relations between material and immaterial entities[72]

- Psychological: Functions related to the judgment, interpretation and appreciation of emotional, behavioral, cognitive, brain functioning, personality, attention, perception, motivation, interpersonal or external environment relationships

- Program: Preset and predetermined policy directives for the initiation, administrating and accomplishment of combined actions requiring contingency by overall operational coordination of the methods and tools. The mind disposes of three programs (perception, verification, response)

- Perception: Program by which the mind scans, recognizes, selects, organizes, comprehends, interprets, records and in general processes and reacts to notions or in general stimuli to which it is exposed

- Verification: Is a program by which the mind can check the conformity of any notion as to a true idea. The mind achieves the verification by its functions according to the level and potential of each level respectively

- Response: Is a program by which the mind can react, respond, translate, link, transmit, diffuse or in general communicate notions to answer and react to achieve action

- Methods: The regular, specific, detailed and systematic procedures needed by the functions to accomplish an action. The mind disposes of three methods (inspiration, processing, and implementation)

- Inspiration: Method by which the mind can capture, figure, formulate, trace, speculate or gain insight, intuition and knowledge of a notion and make it available for processing its information content

- Processing: Method by which the mind puts through the steps of a prescribed procedure through actions to reach a result related to the performance of thinking, memorizing, learning, reasoning and storing knowledge information

---

[72] Wall Street Journal, December 13, 1994.
http://www.udel.edu/educ/gottfredson/reprints/1994WSJmainstream.pdf

- Implementation: Method by which a conclusion of an action is achieved and is put into effect

- Tools: Consist of flexible, investigational, and explorative instruments involved in inspiring, reasoning, planning, learning, communication, perception and the ability to acquire knowledge, move and manipulate material entities, processing and implementing information material as well as immaterial information notions. The tools referred to in the book are grouped respectively to the three methods.

- Instruments: Each tool is operated by specialized instruments. These instruments can be either biological with a cellular infrastructure as with the human mind or artificial intelligent agents and include search, mathematical optimization, logic, probabilistic and other engines providing continuation-based construct that provides timed preemption based on a biological or mechanical clock.

Examples of tools are given as:

- Thinking: Is a tool performed by the instruments of cognition, sentience, consciousness, and imagination.

- Decision: Is a tool performed by the instruments of examination, search of needs, interaction with the environment and rationalizing. The result of the use of the decision instruments is the selection of a course of action among several alternatives. Every decision making process produces a final choice. The output can be an action or an opinion of choice.

- Stimuli: The stimuli are biological bodily signals of subjective information notions projected from the body to the mind centripetally or from the mind to the body centrifugally

- The mind verifies by its various levels (individual, collective, advanced or superlative) the truthfulness of each notion (true and real states, scientific, aesthetic or ethics/activities) as to ideas

- The mind proceeds to the process of all notions even if they are not true as true and real states, scientific, aesthetic or ethics/activities notions. If a thought notion is faulted then the consequent empirical idea and stimuli in the mind will also be faulty. The mind processes thoughts, true or fault, further centripetally to empirical ideas and stimuli

- Empiric ideas: A cogitation material notion built-up by the mind that captures from one hand bodily stimuli and processes them to matrices

- Matrices: A cogitation material notion based on empirical ideas that are in the process of verification as true ideas (true and real states, scientific, aesthetic, and ethics/activities) by any level of mind (individual, collective, advanced or superlative)

- Idea: Is an immaterial, infallible, eternal, immutable, abstract, absolute notion that is provably factual and compliant. The mind processes verified material ideas as thoughts. Ideas exist everywhere outside the human mind. The ideas do not require material entities or energy or any other property or dependence for their existence

- Empedoclean response: Concrete, given and stereotyped response to specific stimuli

INFORMATION SYSTEMS

The transmission of information is through chemical, electric, wave, mechanical or a combination of these. The inbound process is usually a stimulus to receptor transmission, locked by key to key-hold specific affinity that fits securely. Reflexes are common transmission pathways in vertebrates that dispose a central nervous system (Plates 6 to 7).

Information systems are the communication infrastructures used to analyze, study, understand, collect, classify, manipulate, store, retrieve, process, plan, design, test, distribute or support information notions. Such systems are biological information exchanges, cell to cell or intra-cell communications as well as computer science and information technologies and engineering.

Communication in living organisms

The nervous system is a biological information system using a network of information units that transmit signals and coordinate actions between cellular and sub-cellular elements and in higher organisms among tissues, organs and organ systems.

The nervous system is essentially in higher animals an information gateway, responsible for the efferent control of all biological processes and movements in the body and can also receive centripetally information and interpret it. The majority of the information signals are electrochemical, mechanical or physicochemical. In higher animals, including Humans, it is the central nervous system, and the peripheral nervous system, which process the information. The central nervous system consists of the brain and spinal cord. It is responsible for centrifugally receiving and interpreting signals from the peripheral nervous system. The cellular elements of the nervous system are the nerve cells (neurons).

In Table 2 the history of Humans in evolution is presented is the context of the advancement of the information notions processed by the mind. This Table presents the attribution of the information notions to the different levels of the mind. These levels are paralleled to the categories of notions expressed as ideas, which are that are immaterial, infallible, eternal, immutable, abstract, absolute notions provably true scientifically, aesthetically as well as ethics/activity compliant.

The Empedoclean response is defined in this book as a concrete and typecast biological response to specific stimuli and refers to the philosopher Empedocles who first described the effect απορροές as a stimuli signal on the specific πόροι as a receptor thus introducing for the first time in biology the message transmission mode.

.. Moreover, hearing is the result of noise coming from outside... For when (the air) is set in motion by a sound, there is an echo within; for the hearing is as it were a bell echoing within, and the ear he calls an 'offshoot of flesh: and the air when it is set in motion strikes on something hard and makes an echo .. In addition, smell connects with breathing, so those have the keenest smell whose breath moves most quickly; and the strongest odor arises as an effluence from fine and light bodies. But he makes no careful discrimination for taste and touch separately, either how or by what means they take place, except the general statement that sensation takes place by a fitting into the pores; and pleasure is due to likenesses in the elements and in their mixture, and pain of the opposite. In addition, he speaks similarly concerning thought and ignorance: Thinking is by what is like, and not perceiving is by what is unlike, since thought is the same thing as, or something like, sensation. For recounting how we recognize each thing by each, he said at length. Now out of those (elements), all things fit together and their form is fixed, and by these men think and feel pleasure and pain. .. Therefore, it is by blood especially that we think; for in this especially are mingled the elements of things.

TABLE 2 ORGANISMS, RESPONSES AND NOTIONS

ORGANISMS/LEVELS		INFORMATION NOTIONS
Animalia 1	Invertebrates	Empedoclean response*
Animalia 2	Vertebrates[73]	Empedoclean response*

---

[73] Vertebrates, * Empedoclean response is defined in this book as is a reaction to a specific stimulus, as that of an organism or a mechanism. Ideas see Table 3.

Homo sapiens	Early primitive human mind before 350-250.000 years or of a child of the age of 3 to 5 years in 2013 in a developed country		Limited number of real notions proven by trial and error learning, optical recognition (mirror experiment) Comprehension of the notion "Ego sum Ego"
Human 1	Man defined as having a developed and functional mind in a developed country in 2013	Ideas 1	Several scientific notions on the level of the human mind and negligible number of aesthetic, ethics/activities notions
Human 2	Collective mind of a human society with functional mind in 2013 developed country	Ideas 2	Significant number of scientific and a minimal number of aesthetic, ethics/activities notions
Human 3	Collective mind of a multicultural future society	Ideas 3	Numerous scientific notions, advancement in the knowledge of aesthetic, ethics/activities notions
Human 4	Advanced knowledgeable mind of the far future (advanced mind).	Ideas 4	Knowledgeable of nearly all scientific notions and further advancement in the knowledge of aesthetic, ethics/activities notions
Human 5	Absolute knowledge of the totality of all ideas (superlative mind)	Ideas 5	*Logos* absolute knowledge of all aesthetic, ethics/activities notions

This response is a reflective stereotypic answer to a specific stimulus interacting with a receptor that recognizes it and responds to it. A stimuli as a key opens a keyhole only if both fit and recognize each other mutually. The same, a specific receptor recognizes defined stimuli and triggers an exact response.

In Table 3, are defined the various levels of human mind categorizing them to Homo sapiens, human 1 to 5). This Table presents the juxtaposition of notions at each developmental stage in the evolution of living organisms from unicellular species to Humans and the mind.

TABLE 3 INFORMATION SYSTEM OF LIVING ORGANISMS

ORGANISM	INFORMATION SYSTEM LEVEL							
Unicellular	0	0	0	0	0	0	0	0
Multicellular	0	0	0	0	0	0	0	0
Plants	0	0	0	0	0	0	0	0
Animals 1		0	0	0	0	0	0	0
Animals 2			0	0	0	0	0	0
Homo sapiens				0	0	0	0	0
Human 1					0	0	0	0
Human 2						0	0	0
Human 3							0	0
Human 4								0
Human 5								

Vertical column headings: STIMULI-RECEPTOR EMPEDOCLEAN | REFLEXES | CENTRAL NERVOUS SYSTEM | "ego est ego"

Bottom level numbers: 1   2   3   4   5

IDEAS corresponding to levels of the Human mind as shown in Table 2

"EGO SUM EGO"

The notion of "Ego sum Ego" for an individual as member of a society is the recognition of his or her independent and distinct existence as a being separate and distinguishable from the others and from the environment. The creation of the perception of "Ego sum Ego" means the ontological existence of the "self" of the human mind and body. Concretely the notion of "Ego sum Ego" is the necessary consent to consider the mind of an individual as developed, functioning and operational.

The recognition of the notion "Ego sum Ego" is a process by which the mind distinguishes between itself and the "non-self" that is: all that is outside the individual. This notion attributes to the individual the reality of being a distinct entity that is completely separate from all others. As an idea, the notion concretizes the existential appurtenance of the individual and introduces the first notion of consciousness[74] of the mind confirming its being. In addition, this same notion unifies the ontological coexistence of the bodily functions with those of the mind.

Aristotle[75] refers to the ontological terminology stating:

" *But if life itself is good and pleasant … and if one who sees is conscious that he sees, one who hears that he hears, one who walks that he walks and similarly for all the other human activities a faculty is conscious of their exercise, so that whenever we perceive, we are conscious that we perceive, and whenever we think, we are conscious that we think, and to be conscious that we are perceiving or thinking is to be conscious that we exist …*"

By the notion "Ego sum Ego", the perception of self of Aristotle is included into this meaning. "Cogito ergo sum" of Descartes is a notion that comes later in the development of the mind when the self is tested and juxtaposed to its potential existence.

In this book, the notion of "I am I" or "Ego sum Ego" is the first notion processed by the mind that constitutes the borderline of the human mind functions and those of the central nervous system of animals. The central nervous system at this point is jointly functioning in the background, however insufficient to process fully the time-space existential appurtenance of the individual. In contrast, the notion of "cogito ergo sum" has as prerequisites not only the central nervous system to be already functioning but in addition to be consecutive to the notion "Ego sum Ego" as well as to the much later developing complex notions such as "this is not me" or "these are not me", "these are two" and so on.

History of human evolution

Humans 350 to 250 thousand years ago disposed advanced optical recognition and could identify themselves and prove the idea of "Ego sum Ego". In a society of 2013, the corresponding mind level matches to that of a child of 3 to 5 years.

---

[74] Consciousness is the state when the mind is operational and functioning at its best of capacity and is fully alerted and reactive to stimuli.

[75] Nicomachean Ethics, 1170a25 ff.

Referring to Table 2 and 3, we observe a dispersion of the exact time and localization positioning in the ascending pathway in the Darwinian evolution of the systematic categorization of the organisms. The upgrade of Humans to date was gradual and delocalized and followed a heterochronous stepwise development of the mind. The same two Tables set the boundary among the various categories of living organisms as to the biological (genetic) characteristics of each population, the environmental conditions and the level of notions processed.

Human (1) is the average of a Human of today in a developed country disposing a developed and operational mind. In Table 3 are projected the various levels of living organisms from animalia to the human mind on the capacity of verification of ideas.

The notion "Ego sum Ego" sets the separation between *Pan Troglodytes*, other primates, and the primitive human (defined as *Homo sapiens* in Table 3). In the evolution process, some central nervous system functions are varying from species to species and even from individual to individual and possibly some *Pan troglodytes* could develop the notion of "Ego sum Ego" spontaneously and without a follow-up of consecutive complex notions as in human mind.

The projection in time of the Humans as human societies in the future gives a hypothetical level of mind defined as Human 3, 4 and 5; Table 2 presents this in connection to the level of notions at each grade. It is evident historically that human societies after a long and in most instances painful course, nowadays, are approaching an advanced stage of civilization level. Extrapolation to a future scenario is difficult in the context of the fluctuations of the human nature as demonstrated in the past five millennia. The future can vary between a total destruction of human civilization to an idealistic development to Human 1, 2 and 3 levels, as presented in Table 2. The same as in the curve Z of Plate 2 it is speculative any extrapolation to the future. However, in the short-term, the development of the contemporary collective mind is fast and concurrent to the rapid advancement of humanism, science and technology. The subjective trend that appears in the horizon is aiming to a multicultural universal civilization founded on human rights and socialist welfare.

Grading of mind levels

The confirmation of the mind as fully functioning, operational, developed, integrated, vigilant, attentive and alert, conscious and interventional is after it can process the first notion "Ego sum Ego".

By fully "functioning" mind meaning fit, ready, usable, or serviceable for aesthetic, intellectual and psychological functions. By "operational" meaning that, it is effective to a process or series of actions for achieving verification, response and perception of notions. By "developed", meaning that it operates at an optimal level, by "integrated" that it is harmoniously incorporated. By "vigilant and attentive", that it observes and monitors both the external environment as well as the body and by "alert" that it reacts with readiness for action, defense or protection to external dangers. In addition, the mind is "conscious" meaning that it detains power through active Will to achieve active purposes by active actions using exclusively its aesthetic, intellectual and psychological functions. Finally, "interventional" means involvement of the mind as an extraneous or unplanned circumstance, in a state to alter or hinder an action or a development.

The confirmation that the mind is completely functioning is gradual and subsequent to the following three stages:

- The first stage is the optical recognition of self that is through the central nervous system with active Will. In the child, the optical recognition is functional at the age between one and three. The test of the mirror[76] confirms that the child disposes the optical recognition. In parallel, the trial and error learning increases the number of acquired real notions. The consecutive next stage is the inauguration of the mind by the processing of the notion "Ego sum Ego".

- The second step is the self-recognition of the individual's own mind and specifically that he disposes aesthetic, intellectual and psychological functions. The achievement of this is usually at three to five years for a child living in a contemporary western civilization society. In the less advanced societies and in primitive populations it is difficult to determine the age when self-recognition of the mind is functional.

- The third stage is the capacity of the mind to verify notions of science, of aesthetic, ethics/activities; the age when this function starts is difficult to determine. The initiation is differing from person to person and depends on the social, cognitive, cultural and environmental background. The completion of this stage in contemporary societies is at ten to twelve eight; however, examples are that it can start much earlier (Mozart).

---

[76] Ekman P, Davidson RJ, Friesen WV. The Duchenne smile: emotional expression and brain physiology. II. J Pers Soc Psychol. 1990 58:342-53.
http://brainimaging.waisman.wisc.edu/publications/1990/the%20duchenne%20smile.pdf

In this book by definition as a starting point for the complete functioning of the mind coincides when it has achieved at least the first two stages.

HUMAN SOCIETIES

The course of Humans' existence through the millennia is painful, heroic and dramatic. In evolution terms (Plate 1 and 2) the contemporary corresponds to 13.8×10^9 years from the Big Bang and with a temperature in the Universe today of 3 degrees Kelvin.

The genus Homo evolved from the australopithecine to the first *Homo habilis* ancestor and appeared[77] on Earth 2.3×10^6 years ago with some direct or side ancestors of Humans since. As mentioned earlier, Humans as Humans evolved on Earth some 350 to 250.000 years ago as an animal species dotted with mind and therefore differentiated from the other animalia.

Humans gradually acquire the ability to consider the realities from which they depend and further from the everyday routine of survival for nutrition and reproduction now begin to seek the causes that control their existence and try to describe, express and communicate them. This was already established some 30.000 years ago corresponding to the Paleolithic epoch in Africa. It also coincides with the art expression in the caves of France and other places. Humans begin to create science and art and carry out purely aesthetic and intellectual activities, invent tools and produce manufactured objects. They achieve the verification by the mind of some notions of science relative to tools, manufactured objects and their performance based on experience and experimentation. The artistic paintings in the caves of France[78] show that the artist already since then attempts to investigate the underlying truth and cause and verify aesthetic notions. The expression of pure beauty and the social structures show that both aesthetic/artistic, ethics/activities notions are already present in the early settlings of human societies. Humans create science and art using as tools inventiveness and knowledge acquisition to advance toward higher civilization levels and attain verification of science, aesthetic/artistic, ethics/activities notions.

---

[77] The epochs of man's evolution start at the Pleistocene glaciated epoch, to the Paleolithic, Neolithic, and copper, bronze and iron ages.

[78] Chauvet Cave is one of the oldest rock art sites in the world, dating to the Aurignacian period in France, about 30,000-32,000 years ago. The site is in the Pont-d'Arc Valley of Ardèche, France.

With the sharpness and meditative faculty disposed, already at an early date in the evolution, Humans start developing the scientific thought. Science addresses to the few that have the ability to carry out analysis of the causes and connect the observation with definitions and with laws. The scientist of the Paleolithic period or of today uses empirics and experiment based on acquired knowledge and experience to prove the rightness and truth of a process. The verification of a hypothesis based on observation promotes invention that ends with the product. The product of the scientific notions in the human mind might be a stone tool, a spacecraft, a new process, or a theory.

Humans at the same time, then and now, will address to the many through the expression of art. Concurrently in the evolution of Humans' collective mind is the establishment of ethics in a society based on broadly accepted aesthetic/artistic, ethics/activities that will be the basis of moral principles and humanism.

Evolution of societies

Plates 2 and 4 presents the evolution of the Universe based on the interchange of the tetractides of four radicals or elements that pending on the period are a tetrad of quarks, or of elementary particles or of atoms or of nucleic acids. This constant interchange of the basic four elements, since the Big Bang, together with the expansion and the drop of temperature is the background of the evolutionary process of the Universe.

The time elapsed since the formation of Earth is about 5.0×10^9 years during which an ascending evolution is recorded is based on the empirical noocentric observation. The first forms of life as microbes occurred some 2.0×10^9 years later and the first forms of primates appeared before 6.5×10^6 years. The first species of Homo appeared sporadically on Earth before 2.0×10^6 years up to the appearance of Humans.

The evolution is rapid thereafter: art in the caves, immigration of populations, first societies in the Near East, domestication of animals, and invention of the wheel, arboriculture, and agriculture, livestock farming in Asia Minor, Fertile Crescent and first civilizations in Mesopotamia, invention of writing.

The stages of the evolution of Humans are the following periods: apes, nomads, hunting, agriculture and livestock-farming, historical years. The empirical observation that forms the basis of the positive ascending evolution of the Universe are the history records of the advances achieved by Humans in natural sciences, technology, arts, political economy, and sociology among others during this long period. The ascending trend is even faster during the last ten thousands years. The last fifty years in the evolution because of the scientific, technological and socialistic advances, the achievements are revolutionary in society, welfare, education, justice and in several countries are constantly improving. Still, many countries and societies on Earth are slow in progress.

GENETICS AND HUMAN SOCIETIES

Humans and their mind on Earth are under the influence of the genetic infrastructure and the space-time of the environment. To determine the existence of each individual, a zero point genetic instance is set that corresponds to the instance of the biological conception. In most animals and in Humans, this instance coincides with the fusion of the paternal and maternal genetic material to the daughter cell that will form the new individual.

Plate 8 presents the genetic information transferred by the spermatozoid and the ovum to the next generation. Plate 9 presents the biological span of each generation that is under the influence of the respective at each space-time environmental conditions and influence.

Plate 9 describes the arbitrary zero-time start of each individual lifespan, its dependence to the parent genetic information transmitted. The influence of the environment is consistent at all times. The ontological existence includes the bodily cells, tissues, organs and systems as well as the mind that make-up each human as an individual entity. The genetic material in relationship to the information code received from the two parent cells. From this point on, the zero point genetic code is the initial information received and that evolves further during gestation, birth, life to death. This information is a numeral expressed in the four-nucleotide alphabet, and composes a unique code of the individual at the instance of the conception. Further, this code is reproducible under similar conditions anywhere in the Universe and at any time.

Plate 10 presents the phenomenological ascending trend of the Darwinian evolution of species which is under the influences of environmental conditions and the complex interplay of the five variables $\pi 1$, $\pi 2$ and $\pi 3$.

PART 3 METAPHYSICS[79]

The human mind[80], since the beginning of history is the subject of study of philosophy, psychology, sociology, biology, medicine and broadly of humanism. Mythology, theology and the clergies of all religious groups have exhausted all imaginary explanations and have flooded the literature with inconsistencies to describe the emergence of the human race, the human mind or spirit or soul in the evolutionary process of the Universe; however, without scientific basis or proof[81]. The mind as discussed in the next paragraphs is described free of all theological, cryptic, obscure, fanatic, supernatural, mystery or magic dependence.

---

[79] It is not easy to say what metaphysics is. Ancient and medieval philosophers might have said that metaphysics was, like chemistry or astrology, to be defined by its subject matter: metaphysics was the "science" that studied "being as such" or "the first causes of things" or "things that do not change." It is no longer possible to define metaphysics that way, and for two reasons. First, a philosopher who denied the existence of those things that had once been seen as constituting the subject-matter of metaphysics—first causes or unchanging things—would now be considered to be making thereby a metaphysical assertion. Secondly, there are many philosophical problems that are now considered to be metaphysical problems (or at least partly metaphysical problems) that are in no way related to first causes or unchanging things; the problem of free will, for example, or the problem of the mental and the physical. The word "metaphysics" is derived from a collective title of the fourteen books by Aristotle that we currently think of as making up "Aristotle's Metaphysics." Aristotle himself did not know the word. (He had four names for the branch of philosophy that is the subject-matter of Metaphysics: 'first philosophy', 'first science', 'wisdom', and 'theology'.) At least one hundred years after Aristotle's death, an editor of his works (in all probability, Andronicus of Rhodes) entitled those fourteen books *"Ta meta ta phusika"*—"the after the physicals" or "the ones after the physical ones"—, the "physical ones" being the books contained in what we now call Aristotle's Physics. The title was probably meant to warn students of Aristotle's philosophy that they should attempt Metaphysics only after they had mastered "the physical ones," the books about nature or the natural world—that is to say, about change, for change is the defining feature of the natural world. http://plato.stanford.edu/entries/metaphysics/

[80] See in Part 2 in the definitions proposed in this book on the mind

[81] The Greek and Hebrew words that are translated as "mind" are also translated into several other words that seem to indicate that the concept is that of the mind or soul of the person--that is, the inner person, the mind/soul/heart/life/self as opposed to the human spirit or the human body/flesh. The words, mind, soul, heart, life, or self are all synonyms for this part of our 3-part beings. It is the word that is used for a person's will or pleasure, which resides in the soul/mind. It is the life of the person--when the soul/mind leaves that body, the body dies. The brain is there in the body, but the mind has left. http://www.seekfind.net/Mind.html

We propose a novel psychobiology[82] axed on the mind as a hardware information system[83] processing information notions and interrelated to the biological, social, cultural, and environmental factors. This information system is fully integrated to the evolutionary process of the Universe and the Darwinian evolution of species. Taking these considerations altogether, four levels of mind are set: the individual, the collective the advanced and the superlative. The information notions related to the mind are material or immaterial pending on the four levels of mind set beforehand. Furthermore, we put emphasis on the role of genetic heritage and the cultural, technological, social, educational and environmental background.

The prerequisites set are that the mind as described here, exists only in Humans and that it is operational only when all its functions are completely developed, formatted, integrated thus maintaining a vigilant and alert state.

PSYCHOBIOLOGY

The definition of psychobiology given in this book is that it consists of an integrated functional material structure of cells and information notions integrating an organism's body and biological, psychological, and social actions. In Humans additionally it integrates the mind. The mind then integrated embraces the information system that assigns functions, runs programs, applies methods, and operates tools and implements instruments to process information.

This information system is operating through its constitutional anatomical, biological, biochemical and physiological functional infrastructure. In this meaning, the mind is more precisely an information circuit functionally similar to the hardware of a computer with the notions being the software information messages that are processed. The

---

[82]Psychobiology as the interpretation of personality, behavior, and mental illness in terms of responses to interrelated biological, social, cultural, and environmental factors
[83]Is the complex of biological communication faculties that enables verification, response, perception as well as thinking, consciousness, learning, reasoning, action and implementation, characteristic of Humans, but which also in a limited manner to several animals Psychobiology as the interpretation of personality, behavior, and mental illness in terms of responses to interrelated biological, social, cultural, and environmental factors

computer is processing the information through[84] electric circuits and transistors and the human mind using cells, tissues and organs of the body.

The mind processes information as signals that are notions that are reflections of subjective or objective entities as objects, states and phenomena related to the internal ego of the individual or of his environment. The discussion on material notions includes the stimuli, empirical ideas, matrices and thoughts. In this context, ideas are immaterial immutable abstract notions, such as numbers, forms, equations, laws of Science, constants.

The mind as will be analyzed establishes relationships between notions and between a notion and the whole in a set of entities by planning actions to achieve one or more goals under indeterministic conditions of uncertainty.

Plate 11 depicts the proposed Psychobiology and describes the four operational ranks that characterize the human mind operations: functions, programs, methods and tools. The tools are extended to instruments for the exact implementation of the information process. Functions, programs, methods, tools as well as instruments (that are not depicted in the Plate) correspond to the software of computers. The human body in such a paradigm corresponds to the hardware of computers.

Verification of notions

The next chapters will attempt the discussion of the evolution in focusing to the assembly of the mind of Humans and of their societies.

To process the verification of the notion as its conformity to truth and causality[85] and its underlying causality, the mind should be fully functioning, operational, developed, integrated, vigilant, attentive and alert, interventional and consequently conscious[86]. The verification takes

---

[84] Computer: an electronic device using processors designed to accept data, perform prescribed mathematical and logical operations at high speed, and display the results of these operations.

[85] Truth is the accordance, fidelity and conformity of real notions, entities, states, events or snapshots to an absolute objective prototype or idea for which there is categorical, unconditional and total proof. Causality is the relationship between a set of consecutive snapshots.

[86] Conscious as a sense of one's personal or collective identity, including the attitudes, beliefs, and sensitivities held by or considered characteristic of an individual or group. http://www.thefreedictionary.com/consciousness

place for all processing of actions pending on its level, processes repeatedly to the verification of all notions.

In the verification, the mind operates in a systematic pattern of steps through its functions, programs, methods, tools and instruments that follows general archetypes and stereotypes attempting at each instance and for each notion to verify truth and causality. However, each mind develops individual specificities that depend on genetic and environmental conditions.

Consciousness in the context of the definition of mind that is attempted in this book is the state of the fully developed, functional, operational, aware and alerted mind that aims toward active purposes. Similarly, to all voluntary active purposes, consciousness empowerment is by active Will and processes potential actions toward their accomplishment as achieved active actions.

The mind as an exclusive information system of Humans differs from the central nervous system of primates and other higher animalia. However, Humans dispose of an advanced nervous system that is part of the mind as defined here.

Going back to Plate 4 we see that the responses and the consequent actions in relationship to the purposes and the Will and the corresponding determinism and indeterminism of the mind as Humans differentiate Humans from the other animals. As discussed earlier (and depicted in Plate 7), animals as well as Humans disposing a CNS and pyramidic tracts act governed by indeterminism in certain occasions and in this case powered by passive Will.

Nervous system

The development of the mind is gradual and it is a prerequisite that the person disposes an operational central nervous system and that all the body and mind functions are normal corresponding to the respective biological age and the means of each variable that defines the development.

The All-Or-None-Law[87] applies to nerve cell communication by which a nerve (as well as a muscle fiber) does not respond to a stimulus in a

---

[87] Principle stating that the response of a nerve or muscle fiber responds to a stimulus is independent of its intensity. The response is either a complete response or no response with no intermediary graduations.

strength dependent stimulus. Either the response to a stimulus is at strength above a concrete threshold, after which the nerve (or muscle fiber) will give a complete response or it will be irresponsive to it.

A large field of information in the communications between functional entities especially the biological in living organisms is the synaptic affinity between the stimuli and the respective specific receptor. If we focus within the nerve or muscle cell, the electrical stimulus creates the action potential which will occur only when the membrane in stimulated (depolarized) enough so that sodium channels open completely. The minimum stimulus needed to achieve an action potential is called the threshold stimulus and follows therefore the All-Or-None-Law.

In biology, the receptors and many of the stimuli (flux) are biochemical molecules. The stimuli can also be electric pulse or wave, physical, physicochemical or mechanical; the responses are specific for each individual stimuli acting on each individual axis of stimuli-receptor response.

The resulting action from the effect of stimuli on a receptor is a specific response and this interaction works as the key to key-lock effect. In this case, the stimuli as a security token serves as the instrument that operates the receptor, the lock, which will trigger the access to the specific process. The communication hub for the transmittance of an impulse between two or more communicating cells is the synapse where the receptor recognizes the stimuli only if it has affinity to it.

At a synapse several specific chemical transmitters are modulated and in a coordinated manner interact interfering in the electrical signal impulse transmission (that obeys the All-Or-None-Law).

Empedocles as discussed earlier had recognized the significance of the information transmission of optical or auditory signals between the environment and the respective sensory organs, eyes and ears in this case. In his writings he named the stimuli απορροή (flux) and the receptor οπή (hole)[88].

The mind as a communication system is based on the infrastructure of the biological cells and the sub-cellular organelles and the totality of the information notions transmitted. The starting point of every single transmission of all information notions lies on the communications

---

[88] Empedocles described the flux that objects emit that is captured by the specific receptors of sensory organs that are recognized as similar elements that exist inside them. He, in this manner explained the physiology of vision and of audition.

between cells of the body. This is a common factor between Humans and animalia as well as other multicellular organisms. Humans following the Darwinian evolution of species developed specific and characteristic anatomical, physiological and sociological particularities. However, one additional feature is the mind, exclusive to Humans, with functions, programs, methods and tools that are related to the governance of indeterminism and its empowerment by active Will.

RESPONSES

In living organisms starting already from the unicellular ones there are stereotyped responses and reactions when a specific stimuli comes into contact (on a synapse) to a respective specific receptor. The response to such an information recognition transmission is a standard and stereotyped response that ends in the performance of an action. As presented in Tables 2 and 3 responses exist in all organisms and contribute to the intracellular homeostasis and to the homeostatic mechanisms of an organism.

In vertebrates' reflexes are controlled systems, specifically responding to stimuli interlinked through the central nervous system. Plate 12 presents the common domains that some animals share with Humans. In this Plate, the human mind is the differentiating property from animals and even from those that dispose a central nervous system. This is also true even for species that dispose kin recognition (ability to distinguish between close genetic kin and non-kin) but however lack the ability to recognize[89] their-self in a mirror. This ability which is the possession of sense of self is limited to Humans.

Humans and many other living organisms share genetically inherited common properties. The mind as described here is however, an exclusive property of Humans that evolved during the Darwinian evolution of the species from primates. The same Plate shows as the cutting circles the relevance to bilaterally symmetrical animals (chordates, animals with central nervous system, pyramidal and extrapyramidal reflexes). Circles represent the information system infrastructure in all the levels of mind as to the capacity of verification of notions by the different levels of the mind.

---

[89] http://scienceline.org/2011/01/the-monkey-in-the-mirror/

Voluntary and involuntary responses

The hierarchy of responses to stimuli can be voluntary and carried with active actions for an active purpose and be indeterministic as has been discussed earlier and is shown in Plate 7.

Body functions and aesthetic functions of the mind

The process of the aesthetic mind functions is to verify the notions as to their underlying beauty (exquisiteness, attractiveness, magnificence, loveliness, exquisiteness, splendor and more). Both body sensory and mind aesthetic functions capture, interlink, transmit and distribute the information however only the latter process the analytical and synthetic tasks of verification and proof of the truth and causality.

Plate 6 shows the informatics of the body and of the mind based on the communication systems and functions involved in each case. The mind uses as its infrastructure the whole body (cells, tissues, organs) supporting all the mind functions (aesthetic, intellectual and psychological). The cells of the body and in particular the sensorial cells and the central nervous system operate both the aesthetic mind and the body sensorial functions.

The aesthetic functions of the mind refer to the mind notions related to the nature of beauty, art, taste and appreciation in general of the person's qualities and skills in the context of the environment that are assessing the sensory judgments. The examination, evaluation and appreciation of an object and even more of a work of art are the abilities to form at least a subjective opinion about it. The appreciation is pending on the level of the mind when taking into consideration the knowledge, background, and understanding the objective, unquestionable qualities that comprise the object or the work of art.

The infrastructures of the body sensory functions are the sensory receptors, the neural pathways, and in particular, the central nervous system functions involved in sensory perception. Vision, hearing, somatic sensation, taste and olfaction comprise the sensory system.

The body sensorial systems are transducers from the environment of all the subjective information notions that are projected and made available to both the rest of the body and the mind. The information transduced by the sensorial system elaborates, evaluates, assesses and define the aesthetic function of the mind. Thereafter, together with the intellectual and psychological functions, produces processes and interacts the mind material notions (thoughts, matrices, and empirical ideas).

The body sensorial systems interact as well through material information notions, the stimuli that are also perceivable by the mind functions after their transduction to empirical ideas.

Animals, mind and instincts

Animals do not dispose a mind, however, in the higher vertebrates, central nervous system functions and in particular, reflexes operate through pyramidal (voluntary) neurons that are also common to Humans.

Animals, like Humans, dispose of powerful sensorial systems that in some species are clearly superior to the corresponding sensorial organs of Humans. Both superior animals and Humans dispose of instincts that are stereotyped reflexive responses toward specific stimuli. The instincts as inclinations are variable from one individual to the other owing to different genetic, learning, training, experience and environmental conditions.

Passive Will drives responses to instincts. Instincts, as standardized processes of behavior toward concrete stimuli, are the motivating passive purposes to act and to respond to biological needs (like survival and perpetuation of species).

Instincts are rooted in each individual or in a society and expressed as impetuses, social taboos, or religious ceremonials. It is difficult to define the limits of normal to a physiological instinct and the boundary between instinct and impetus. The biases, the moral barriers, the commissioned discipline, the fanaticism in the patriotism and the religion are stereotyped responses in stimuli that are forced, imposed and adopted by family, society, nation or race to a person or to a society. Plate 7 shows such responses to acquired religious or social moral taboos that end-up becoming stereotyped passive actions similar to instincts.

Compulsively and purposely imposed stimuli, often in a fearsomely and interventional manner cause impetuses. Active Will seemingly only carries out such impetuses as active purposes. In fact, such impetuses are passive and are the result of long impositions of social, religious, nationalistic, ethnic, and militaristic or sect ethics, traditions and customs. These forced impetuses in Humans when pertained for long periods covering many generations end-up as physiological instincts like in animals.

MIND: INFRASTRUCTURE

Each cell, tissue, organ or organ system of the human body is a part of the mind as well as all the information that is processed or that is stored in each individual. The human mind is a communication information system processing information notions; it is automated, electrochemical, and reprogrammable as a system with circuits of signals that are processing information. The mind processing operates through predetermined functions and programs that operate with particular methods and tools.

The human cells and the sub-cellular components and organelles are the common infrastructure of both the mind and the body as defined in this book. This definition differs from the classical biological approach considering as the primary role in thinking and in general mental functions the nervous system[90] in general. It also differs from philosophical schools defining the human mind as separate from the body.

Every Human individual in a society disposes an individual mind that is distinct from the collective mind of the society where he lives and belongs to, and operates as a material integrated entity. The basic elements of the material infrastructure of the mind are from one hand the cells, the tissues and the organs of each Human and from the other the information as message signals. Therefore, the mind apart from its material cellular infrastructure disposes a dynamic information system analogous to the software of computers.

The specific characteristics of each individual mind is differentiated following the evolution responding to the context of the biological, bodily, social, cultural and environmental stimuli exposed. Therefore, the mind is the decision-maker and executive instrument of all senses, intellect and sentiments by initiating potential actions (in the preparative planning stage) and accomplishing them (praxis).

Start of mind functioning

The jump from the animals including the closely related to Humans primates to the mind is gradual and therefore cannot be defined as a "one go" single stroke event. The description of the evolution that occurred in stages detailed as Human 1, 2, 3 and 4 in Tables 2 and 3.

The use of the programs of perception, verification and response initiate the gradual start of the mind. Tables 2 and 3 present the precocious recognition of the self, as the initial, signal for the start of the human mind. The stage of the mind as "Human 1" (Table 2) is a mind of a human in a developed country in 2013 being developed and functional

---

[90] Ancient Greek authors wrongly believed that the heart plays this role.

and having the ability to verify several scientific notions including the comprehension of the notion "Ego sum Ego".

The final start of the functioning of the mind is the process of the verb "essere" which becomes perceptible, combined with the pronoun "ego" which internalizes the existential unity of the individual. The elucidation of the first notion will create a second notion of whatever is outside me is "not-me". The two notions "Ego sum Ego" and "not-me" direct to the creation of a new notion: the number "two". At this point, the mind is operational, responds to the criteria set (Table 2, Human 1) and is ready to operate upgrading with each new notion captured and processed.

The gradual Human 1 and 2 levels are the criteria that refer to the attainment of the mind to the average performance of the Humans of today in a developed society. Further, they refer to the accomplishment of the evolutionary stages to Humans and that the mind has already the capacity of verification of several scientific notions to proved ideas.

The first notion of "Ego sum Ego" includes the coordinate maturity of the body functions (Plate 13) and the mind functions. The maturation of the mind's functions, programs, methods and tools is attained gradually up to the point when there is a comprehension of the notion "Ego sum Ego" (Table 3). At this point by the definition given in this book, the Human 1 and 2 levels corresponds to the average of an individual living in a contemporary modern society.

Table 2 and 3 details the variable conditions of age in the maturation of the mind; these depend on the societal, technological and environmental conditions.

Mind level and societal civilization

A large variability among human societies renders the criteria set in Tables 2 and 3 arbitrary and only indicative. The variability relays to the historical time the conditions of each individual and society, the particularities of individuals and many others. Therefore, the attainment of the "Ego sum Ego" first notion timing is also variable.

Plate 11 presents the architectural structure of the mind and the cascade of the Psychobiology as described and defined below.

The aesthetic and intellectual functions of Humans developed very rapidly in the last thirty thousand years starting with the cave paintings in France and elsewhere. Simultaneously there is a more latent development of the psychological functions. Sporadically several higher animals have developed functions that are advanced and comparable or even occasionally more significant than in Humans. There is evidence that some species of animals share rudimentary psychological operations[91] resembling human societal rituals.

Going to Table 2, Human levels 3, 4 and 5 are projections of the human societies on Earth based on hypothetical considerations stating that the evolutionary ascending trend observed to date continues with no major perturbations. The technological development, the educational upgrade of the world population, the globalization of knowledge and information transfer theoretically leads to a mature multicultural future civilization. Such projections are hypothetical and liable to opposite endpoints of retrograde or even destruction in case of major changes in the scenarios. Changes are possible due to the unpredictability of the human nature and behavior or to major physical or environmental perturbations.

BODY AND MIND FUNCTIONS

The body and the mind form the two modules of Humans and this statement is referring to Aristotle's[92] postulation that the body and the mind exist as facets of the same being, with the mind being one of the body's functions. As to the "nous" he states that it consists of two parts: something similar to matter (passive intellect) and something similar to form (active intellect).

In the Aristotelian context, the positions held in this book are that the human mind is material; however, its objects are either immaterial or

---

[91] Chimpanzees have been reported to have funeral rites and to take away the bodies of the deceased after death

[92] Aristotle says that mind (nous, intellect) is separable, impassible, unmixed, since it is in its essential nature activity: "When intellect is set free from its present conditions, it appears as just what it is and nothing more: it alone is immortal and eternal and without it nothing thinks". After characterizing the mind (nous) and its activities in De Anima iii 4, Aristotle takes a surprising turn. In De Anima iii 5, he introduces an obscure and hotly disputed subject: the active mind or active intellect (nous poiêtikos). Controversy surrounds almost every aspect of De Anima iii 5, not least because in it Aristotle characterizes the active mind—a topic mentioned nowhere else in his entire corpus—as separate and unaffected and unmixed, being in its essence actuality (chôristos kai apathês kai amigês, tê(i) ousia(i) energeia; DA iii 5, 430a17–18) and then also as 'deathless and everlasting' (athanaton kai aidion; DA iii 5, 430a23.
http://www.archive.org/details/aristotledeanima005947mbp

material. The position put forward is that the material entities, including the mind, are compatible with the immaterial notions: the mind can verify the material notions. Further, the position held is that the active Will, powers the human mind; the central nervous system of Humans, primates and of some higher animals is powered by passive Will. Respectively there is a separation of the purposes to active and passive and of resulting actions to active and passive.

Architecture

The mind being globally material entity has a distinct anatomical and physiological hardware. Humans and several higher species of animals, share this part of the hardware. The object, of both Humans and these animals, is the processing of material information notions. However, the mind has the exclusive ability to verify as well material notions as to their equivalence to immaterial ideas and attempt to prove the underlying truth. This ability is related to the level o the mind.

Plate 11 shows this structural hierarchy for the processing the notions with the use of the four operational ranks (functions, programs, methods, tools) that distinguish the mind operations.

A notion is a signal information message that contains all the essential data so when captured and restored, reproduces all the contained initially information. Immaterial notions are abstract information: like numbers, digits, chemical formulas, and laboratory results, configurations of solids and volumes, mathematical relations, factors, constants, ranks of mathematics, geometric forms, musical tones, and colors among other. Here, we define as ideas only notions that are true, infallible, scientifically verified, proven and real. Globally all ideas are included in the *logos* which is a finite set. The total number of ideas encompassed in the *logos* is a natural number

The notions of the mind are material and include thoughts, matrices, empirical ideas and stimuli. The operation of these notions is through signals that are usually electrochemical and frequently mechanical, acoustic, optical, olfactory, tactile, stimuli or waves. The functions of the mind are the aesthetic, intellectual and psychological.

The human mind information processing architecture is depicted in Plate 11 while in Plates 13 and 14 are shown the relations to the environment and to the immaterial notions. Plate 13 in particular shows the functions of the body and the mind as to the environmental and physiological stimuli that bridge the communication between the environment from one hand and the body and mind from the other. Plate 14 distinguishes the material notions and the immaterial as to the mind and the body and shows the compatibility of the ideas to all the other material notions despite that the integrity of both mind and body are material.

Informatics of mind and body

Plate 6 depicts the informatics of the body and mind. The two parallel columns show the respective functions, programs and methods respectively for the mind and the body.

The body functions are the sensory, respiratory, cardiovascular, urogenital, digestive, nervous system etc. All the body homeostatic mechanisms, the contractions of muscles, the induction of stimuli, the synergistic or antagonistic actions of receptors and many others are programs for the performance of body functions. Tools are in the body all the receptors and the totality of the electrochemical and biochemical reactions. On the left column of the same Plate 6 are the respective functions, programs and methods related to the mind.

In Plate 13, the purely bodily functions are shown in white background shading while the purely mind functions in gray. Both the mind and the body modules use the same histological, anatomical and physiological infrastructures (cells, tissues, organs, organ systems). The ideas, as immaterial notions are in a black-background. The stimuli in this plate are presented in white and grey triangles.

The stimuli (Plate 13) participate in the communication among cells, tissues and organs and comprise an internal bodily or external environmental set of information that is can trigger a response. The stimuli operate through purely bodily functions and are the platform by which stimuli and empirical ideas are interacting and processing information. The bridge of the information of the human between purely bodily and mind communications are the empirical ideas that will interact with all the processing of the mind (method, programs, functions).

Functions

Going back to Plate 6 we show the differences and the analogies existing between the body and mind modules of Humans. The aesthetic functions of the mind refer to the analysis, synthesis, rationalization, systematization and finally verify the cause and the truth underlying the sensorial information captured by the bodily sensory functions. The binary appurtenance of the body cells to both the sensorial (bodily) and the aesthetic (mind) functions allows the continuous communication between the two modules.

The intellectual function follow linear retrospective correlations with constant factors[93], on the contrary the aesthetic and psychological follow nonlinear multiple objective interrelations with continuous variables of control[94].

The intellectual functions refer to the objective understanding and are opposed to the subjective emotions. They refer to the ability to objectively rationalize avoiding emotional sensual and mood related interpretations with proved supportive explanations. The intellectual functions include Plato's[95] *dianoia* and *noesis*. Aristotle states on this subject *"in human beings it is the rational soul[96] that is capable of thought and reflection"* (differentiating it from the the other two psyches the vegetal and the animal).

The psychological functions refer to the emotional and affective responses and the activation of the expressive spheres of the central nervous system and in particular, those related to the behavior. The Primates and superior animals share responses to stimuli with Humans, for example interpreting and translating an olfactory stimulus with a passive action. The human mind will differ by responding to the smell of a pheromone not only by passive instinctive sexual arouse but also by an amorous psychological verbal expression of his or her sentiments which is a mind function.

Programs

The mind as defined earlier[97] disposes of three programs (perception, verification and response). The purpose of these programs is to elaborate and operationally coordinate the notions in order to be further processed

---

[93] Function in which there is a constant change in respect to the input. The graph of linear relations is always a line. The general format for a linear relation is: $y = ax + c$ where, "a" and "c" that are constants.

[94] Is a nonlinear system that is, a system which does not satisfy the superposition principle, or whose output is not directly proportional to its input.

[95] Plato's Analogy of the divided line, noesis is beyond dianoia, and while dianoia concerns mathematical entities, noesis is the highest state of the mind which reasons from Forms to Forms, reaches first principles, first causes and the axioms of axiology, and deduces conclusions from them.

[96] Aristotle described the soul (psyche) as a substance can receive knowledge and that knowledge is obtained through the soul's capability of intelligence, although the five senses are also necessary to obtain knowledge.

[97] See Part 2, Chapter Definitions

through the methods and tools. The programs accomplish combined actions requiring contingency by overall logical analysis and synthesis (following the Aristotelian logic[98]) for the deduction (συλλογισμός) of each notion as to its content and its relationship to other notions (Plate 11).

The program of perception corresponds to the Aristotelian analytic establishment of the exact premises of each notion (πρότασις) by enclosing all notions and not limited to the speech notions. The verification program corresponds to the Aristotelian induction (επαγωγή) that based on the analysis of the premises proceeds to the synthesis verification of the notion. The response program corresponds to the Aristotelian conclusion (συμπέρασμα) thus completing the syllogistic deduction (συλλογισμός) and presenting the notion to the methods and tools.

Methods and tools

The tools are investigating and implementing the perception, verification and response programs. The achievement of this is through the methods of inspiration, processing and implementation. Inspiration is the stimulation of all the mind functions to achieve a higher-level notion aiming to novelty, creativity, new knowledge, improved process or performance of quality. The processing program consists of a series of intermediate preparatory operations, actions, changes, or functions that aim in achieving a predetermined endpoint. Implementation is the performing and rendering effective an ongoing process and put into effect.

The tools will utilize further instruments according to each of the three methods examples of which are as follows:

- For the first method of inspiration tools/instruments are: research/searching, studying, differentiating, exploring, probing, investigating, stochastic/guessing, supposing, hypothesizing, assuming, speculating, devising, aiming, targeting, randomizing, sequencing, deducing, presuming, scanning, selection/assorting, ranging, comparing, estimating, collecting and harmonization/coordinating, organizing, rationalizing, managing, synchronizing, assembling.

- For the second method of processing tools/instruments are thinking/believing, reasoning, formulating, supposing, memory/reminiscence, recollection, remembrance, knowledge/learning,

---

[98] Organon: http://plato.stanford.edu/entries/aristotle-logic/

being conscious or aware of, understanding, comprehending, informing, familiarizing, apprehending and archiving/databank, listing, indexing, registration, compiling.

- Last, for the third method of implementation tools/instruments are: decision/concluding, judging, assessing, resulting, resolving, mobilization/recruiting, listing, recording, attachment, intervention/involving, enfolding, enveloping, encumbering, intersecting, entrusting, relating, achievement/attaining, accomplishing, completing, realizing.

These tools/instruments are exhaustive and can be synonymous or extended to other similar that however, altogether are involved in processing information notions.

The tools that operate in the mind are powered by active Will and are under the governance of indeterminism respectively. Randomness in the macro Universe of the biosphere where Humans are is negligible, however the interplay of the five variables enable randomicity in the microcosmos.

OPINION AND KNOWLEDGE

The mind is a functional information system ruled and complying with the laws of Science, to the environment and in general to the Universe. We propose a novel information system, in the context of the ontology of the mind juxtaposed to idealism and materialism.

The mind as an information system is composed of message signals described as stimuli, empirical ideas, matrices, thoughts that are material and it is compatible to immaterial ideas.

Ideas, using the Platonic definitions, are the principal reality in contrast to the individual objects of sensory stimuli (experiences) that undergo constant change and flux. Plato held that ideas are perfect, eternal, and immutable, and consequently knowledge of material things is not real knowledge. In this sense, the verification of a notion confirms that the information attached to it is true. Therefore, the information contained in a verified and proven as idea notion are eternal, abstract and unchangeable. Following Plato, notions processed by the mind are "opinions" and only the verified and provably are true "knowledge". The mind investigates the material stimuli, empirical ideas, matrices and thoughts processing them for verification to ideas. The verified material notions when proven and found true, then become ideas; if not and found false, or unproven then they remain material notions as subjective opinions

Levels

The mind lies between ranges of levels that describe its aptitude to verify the material notions. The levels described are the individual mind of a contemporary man in a developed society, the collective mind of a modern human society, a future advanced mind and finally the superlative mind that can verify all notions and therefore disposes the knowledge of all ideas. The totality of all the ideas is the *logos*.

Plates 15 and 16 depict the material notions of the mind (thoughts, matrices, empirical ideas and stimuli) which potentially are verifiable to correspond to ideas. In Plate 15, random numbers despite that are as abstract immaterial notions they are not ideas since they do not correspond to a concrete notion. Errors and fantasies of the mind that are material *de facto* do not correspond to an idea.

The individual mind that is the lowest mind level executes aesthetic, intellectual and psychological functions. To accomplish these functions the mind is already operational conscious, alert, vigilant, attentive and watchful. Notions that are in the unconscious as well as oneiric, phantasm, hallucinating, delusion, chimera and illusion notions are processed but are not real or liable to constitute ideas. Plate 16 depicts the flow of stimuli to empiric ideas and matrix by observing a real object (in this case the statue of Venus of Melos).

NATURE OF NOTIONS

The human mind notions are material and include stimuli, empirical ideas, thoughts and matrices. In contrast, immaterial notions while existing as ideas, are independent, compatible to but not part of the human mind. The mind aims in the conformity of each notion as to an idea: its target is the verification and proof of the true of all material notions. The absolute target is that the mind can process the totality of the existing ideas, this meaning the *logos*. It is only by knowledge, experience, experiment and research that the mind increases its repertoire for conformity of proven material notions to immaterial ideas.

Plate 14 presents information pathways in the mind leading to the ideas and by extension the *logos* being immaterial (on the left in dotted lines) in contrast to all the material (on the right) in continuous lines. This Plate distinguishes between the material and immaterial notions as well as the various ranks of the two modules of the Humans. The same Plate shows the frontiers of the notions between[99] idealism and materialism.

Immaterial notions

Plate 15 depicts the notions processed by the corresponding levels of the mind. The immaterial information notions are abstract that while free of energy, space, time or any other dependence, still can with high fidelity, exactitude, accuracy and precision be transduced and converted to material digitized[100] signals. Vice versa, the respective digital signal transduced and converted back to the initial immaterial notion. This back and forth circuit of verification or rejection is processed accordingly by the different levels of the mind.

The same Plate 15 shows the notions, contained in touching circles, the immaterial notions (ideas) depicted in gray. The Plate presents the ideas in three circles pending on the level of the mind with the smaller circle depicting the individual mind and the larger the totality of ideas (*logos*) of the superlative mind. Among the immaterial notions are included all the numbers both those representing concrete ideas and those that are random "nonsense" that do not correspond to any concrete notion.

Ideas as mathematical digital numerals

The underlying common denominator of all ideas is a digital numeral that describes the idea about its full content, description and properties. The digital numeral is the transformation equivalent of the idea. This numeral is reproducible to a concrete material respective entity. The achievement

---

[99] Idealism is the philosophical theory, which maintains that the ultimate nature of reality is based on the mind or ideas. The theory of materialism holds that the only thing that exists is matter; that all things are composed of material. In other words, matter is the only substance. As a theory, materialism is a form of physicals and belongs to the class of monist ontology. The duality between the body and the mind remains a field of thorough discussion since Plato and Aristotle, through Descartes until our time. Dualism and monism are in this respect the two main lines of thought. The further division of dualism refers to the substance and to the property dualism. On the other hand, monism holds that the mind and the body are the same ontological entities. Another way of interpreting the mind and body functions is by arguing that only material entities exist (physicalists) or on the contrary that the mind is the only real entity and all the rest are illusionary creations.

[100] A digital system is a data technology that uses discrete (discontinuous) values. By contrast, non-digital (or analog) systems use a continuous range of values to represent information. Although digital representations are discrete, the information represented can be either discrete, such as numbers, letters or icons, or continuous, such as sounds, images, and other measurements of continuous systems.

of this is by the capture and transduction of the idea to a respective material entity under appropriate circumstances.

A number, as abstract notion, is a set of positive integers from a series of symbols of unique meaning that is in a fixed order derived by enumerating, or listing or detailing. However, not all numbers are necessarily ideas; numbers that do not correspond to a concrete idea are random "nonsense" numbers. A "nonsense" number is an abstract notion however, not corresponding to a concrete idea and is therefore, not included in the *logos*.

Other abstract notions are the mathematical and physical constants. A physical constant is a physical quantity that is both universal in nature and constant in time. In contrast, a mathematical constant is a fixed numerical value that does not directly involve any physical measurement.

Immaterial notions include all the digital numerals corresponding exactly to the form of whatever material or construction anywhere in the Universe. It also includes all the genetic codes of all living organisms. The corresponding digital numeral describes its shape, appearance, countenance, quality and all its properties in complete detail. This means that the initial material or construction is exactly reproducible when a transducer captures their description.

Material notions

Material notions are all information signals expressed in any energy or matter dependent manner. Material notions are electric, chemical, mechanical, wave, radiation energy or any other transfer transmitted as a signal that when captured results to a concrete response. The material notions of the mind are the thoughts, matrices, empirical ideas and stimuli that are mainly electrochemical and mechanical (Plate 13).

Categories of notions

The categories of notions are three: those referring to real states and science, those to aesthetics/artistic ones and finally those to ethics/activities. The basis of the evaluation of the achievements of the human societies on Earth is by the assessment of the scientific, aesthetic/artistic and ethics/activities notions that globally classify the civilization level.

The mind for each category of notions uses all three functions (aesthetic, intellectual, psychological). However, the process of scientific notions is mostly by the intellectual functions, of the aesthetic/artistic is mainly by use of aesthetic functions and of ethics/activities by psychological.

In each notion category, several factors interfere. Especially in the scientific notions, the level of education and the technological infrastructure of research are playing a major role. In general, the civilization and socioeconomic background combined with the relevant history of each epoch influence all three categories of notions. The assessment by using labels as "proven", "true", "good" or "beautiful, justice, virtuous, wise, prudent" or synonyms, variables, or opposites renders the categorization of the notions highly complex.

The mind verifies and processes the notions as true and real or scientifically proven, beautiful, justice, virtuous, wise and prudent. Numerous other synonyms, equivalents, or similar are processed by the same mind mechanisms.

Parmenidian εἶναι and ideas

Parmenides of Elea, active in the earlier part of the 5th century before our era, composed a difficult metaphysical poem that has earned him a reputation as early Greek philosophy's most profound and challenging thinker.

Parmenides[101] in his poem on the real existence of the being (εἶναι) says:

"Come now, I shall tell and convey home the tale once you have heard just which ways of inquiry alone there are for thinking: The one, that [it] is (εἶναι) and that [it] is not "not to be", is the path of conviction, for it attends upon true reality. But the other, that [it] is not and that [it] must not be, this, I tell you, is a path wholly without report: for neither could you apprehend what is not, for it is not to be accomplished, nor could you indicate it. It is necessary to say and to think that what is *is* (εἶναι); for it is to be, but nothing it is not. These things I bid you ponder. For I shall begin for you from this first way of inquiry, then yet again from that along which mortals who know nothing wander two-headed: for haplessness in their breasts direct wandering thought. They are borne along deaf and blind at once, bedazzled, undiscriminating hordes, who have supposed that it is and is not the same and not the same; but the path of all these turns back on itself".

---

[101] http://Plato.stanford.edu/entries/parmenides/ and
http://www.ellopos.net/elpenor/greek-texts/ancient-greece/parmenides-being.asp

True and real material entities in contemporary terms refer to their proven existence at a concrete space-time that is accorded to a precise snapshot. Whatever is self evident and is objective is a true and real material entity; for example if we see the moon as a complete material entity with our eyes and everybody in the human community sees also the moon as such then the moon is a proven notion and is εἶναι. This does not mean that the details of the map of the moon macroscopically are yet proven. However, when the telescopes, the satellites and the exact mapping of the moon were achieved by science then the depicted geographical localities of the moon became proven scientific notions. With this example we show that the moon in earlier historical times was objectively a real material entity and the notion of the "moon" as observed by the human society of the time was a proven idea. In our times that we have scientific and experimental proof of the moon valleys, craters and mountains these localities become proven ideas as well and are (εἶναι).

Equivalence

True and real also refer to the absolute equivalence and reproducibility between two entities that have the property to be transformed from one expression of its ontological existence to the other in due space-time. Such inter transforming entities are those between a virtual and real material expression. Parmenides is chiefly interested to prove that (εἶναι) *is;* but it is not obvious at first sight what it is precisely that *is* (εἶναι). The introduction of the scientific and experimental proof for the true εἶναι clarifies this point.

Focusing on true and real entities that provably exist we take another example: an iron rod of specified dimensions and other properties. This rod is material fully described in virtual digital concrete information. This information is sufficient under the same conditions that simulate the conditions of the original rod to reproduce an identical rod. The rods differ is time and space otherwise are exact replicas. In this sense, the material entity (rod) is information wise equivalent to the digital numeral describing it and allows the reproduction of the rod.

"One[102] path only is left for us to speak of, namely, that *It is*. In it are very many tokens that what is *is* uncreated and indestructible; for it is complete, immovable, and without end. Nor was it ever, nor will it be; for

---

[102] http://lexundria.com/parm_frag/1-19/b

now it is, all at once, a continuous one. For what kinds of origin for it wilt thou look for? In what way and from what source could it have drawn its increase? I shall not let thee say nor think that it came from what is not; for it can neither be thought nor uttered that anything is not. And, if it came from nothing, what need could have made it arise later rather than sooner? Therefore must it either be altogether or be not at all. Nor will the force of truth suffer aught to arise besides itself from that which is not. Wherefore, Justice doth not loose her fetters and let anything come into being or pass away, but holds it fast. Our judgment thereon depends on this: "*Is it* or *is it not?*" Surely it is adjudged, as it needs must be, that we are to set aside the one way as unthinkable and nameless (for it is no true way), and that the other path is real and true. How, then, can what *is* be going to be in the future? Or how could it come into being? If it came into being, it is not; nor is it if it is going to be in the future. Thus is becoming extinguished and passing away not to be heard of.

Nor is it divisible, since it is all alike, and there is no more of it in one place than in another, to hinder it from holding together, nor less of it, but everything is full of what is. Wherefore it is wholly continuous; for what is, is in contact with what is."

By juxtaposition, the Parmidean reference on the impossibility of an entity that is (εἶναι) to have birth or death (since it is in all its entity now) the equivalence of the rod exemplifies this statement. The material rod since real is equivalent to its respective immaterial idea that is a digital numerical. This is reproducible and can again become the material rod: therefore, the rod either real and material or virtual and immaterial never has birth or death since the ideas are immaterial and eternal.

Parmenides goes on to consider in the light of this principle the consequences of saying that an entity *is* (εἶναι) and cannot not to be.

This, is because it cannot have come into being. If it had, it must have arisen from nothing or from something. It cannot have arisen from nothing; for there is no nothing. It cannot have arisen from something; for here are only existing entities that are (εἶναι). Nor can anything else besides itself come into being; for there can be no empty space in which it could do so. Is it or is it not? If it is (εἶναι), then it is now, all at once.

In this way, Parmenides refutes all accounts of the origin of the world *Ex nihilo nihil fit.*

Going back to the example of the iron rod, the rod exists always both ways, either material (iron rod entity) or immaterial (virtual). In a following chapter, we describe a related example on equivalence is described referring to the equivalence of a living organism (the influenza virus).

True (real) state notions

The categories of the notions, as said earlier, are the following: true (real) states, scientific, aesthetic and ethics/activities.

The true (real) states are those that objectively occur and proven scientifically as true. These notions describe, as explained in the example of the moon, the topographical, chronological and ontological reality of a notion in a way as a photographic objective snapshot. The material object described by such a notion in the Universe effectively and objectively exists and thus is true and real.

The verification by the mind has to be unquestionable especially since it is based on sensory bodily functions. The mind, for the verification of a true (real) state, uses as infrastructure the body functions (Plate 6). However, the body and in particular the sensorial organs functions (including the esthetic) despite that they are anatomically the same, operate independently from the mind functions.

As an example, when the mind records a rose as a flower that is seen as such and can be experimentally photographed and analyzed then even if the rose is perceived by defiant human senses still it is a true material object. This rose as a generic is recognized within a concrete topographic locality and at a specified time and its state is determined as to the other surrounding objects and states. In this case, this generic rose is an unquestionable and proven notion for its existence and it is real and true. The rose for the observer in such a case its existence is not a mere opinion but is true knowledge.

Still, the observer does not know all the properties of this rose; for example, the genetic code or the biochemical properties of the rose in question are unproven or unknown. Therefore only the generic of the rose as seen, recognized and identified is a real notion the rest of the properties that are unproven or questionable are unverified notions. All the other notions describing the rose either being scientific or aesthetic or ethic/state remain to be verified.

The same is with the example of the moon: the as a generic is a real notion, however the topography of the moon till recently was not proven scientific notion but all descriptions were speculative. The same is valid for the, the sky, the sea, the Earth, a mountain and so forth.

Venus of Melos true (real) situation notions

Plate 16 presents an example with the statue of Venus of Melos that is in the Louvre and shows interchanging transformation between a generic notion classified as "statue" and a specific notion "Venus of Melos in the Louvre Museum". Both the real material entity and the information notion are true, scientifically, and experimentally proven since the statue exists and it is a concrete sculpture creation of "Venus of Melos in the Louvre Museum" by photography and tomography scanned and proven. Therefore following the Parmenidean definitions both notions: the "statue" and "Venus of Melos" εῖναι and truly exist. That is, the first as a generic true (real) situation, the second as a scientific proven notion.

The observer of the real statue (presented with an eye at the top right of the Plate) when he perceives the statue standing in front of him at the Louvre forms the unquestionable notion of the real and true statue that he is observing. This statue, in front of the observer, is real, true, and labeled as "Venus of Melos". At the best of the knowledge of the observer, in front of him is a statue and this statue is named "Venus of Melos". Therefore, the notion of "statue" is a real notion for the observer and constitutes a proven idea.

However, the notion "Venus of Melos" for the naïve observer is an unproven scientific notion that subjectively needs verification for being an idea. Now if the observer is a contemporary archaeologist who has studied the discovery of the statue, has checked the photographs and other experimental proofs then for him or her "Statue of Melos" is a scientifically proven verified idea.

Going into more details, in the mind of either one of the two the observers, the real statue will create mind stimuli (a) which will further be processed as an empirical idea which will in turn be assembled as a matrix (b). The matrix in the mind will be processed for verification and will be projected as a thought. The thought will re-enter the cycle and will create a modified empirical idea and so on. The stimuli, empirical idea, matrix and thought are subjective unproven notions that require verification to the respective idea (f). If the notion of matrix is verified and proved as in the case of the archaeologist then the "statute" and "Venus of Melos" notions are by now ideas, otherwise these notions remain unproven pending notions. The notions that have been proven as ideas are immaterial and on the contrary the all the other mentioned notions (a, b, c, d and e) are material. The mind however even by verifying the notions as true (real) and scientifically proven ideas, still remains material in its totality.

Now, if we have a third observer who is not a direct observer of the statue but only a reader of a post-card or of a book describing "Venus of Melos" the proof of the formed empirical ideas and matrices is more complex. The proximity of the direct observer renders easier the verification of the material notion at least of the notion "statue" as a true (real) notion and therefore a provable idea. In Plate 16 the pathways f and g correspond to the immaterial notions related to the true (real) and scientific notions of the "statue" and "Venus of Melos" respectively.

Referring to the observer, the archaeologist or to the reader, we assume that both dispose of a contemporary mind that relates to the collective mind[103].

"Venus of Melos" other notions

The observer's mind has the capacity as discussed in the previous paragraph, to verify the notion of "Venus of Melos" statue standing in front of him as effectively an idea. However, behind the real situation of the statue and the scientific notion of its identity, underlie aesthetic, intellectual and ethics/activities notions. Questions related to the beauty of the statue, or of its proportions, expression, causal drive of the sculpture, exhibitionism, sexuality, morality among others remain as material matrices and empirical ideas in the subjective opinion of the mind of the observer. However, all these questions correspond to proven ideas based on knowledge that the level of mind of the observer cannot verify at this point. The pathways f and g correspond to the immaterial notions related to these aesthetic, scientific, and intellectual and ethics/activities notions.

We have used the example of "Venus of Melos" to differentiate the respective notion categories.

- The observer perceives the image of the statue through the sensorial functions (bodily sensorial stimuli).

- As discussed the observer pending on his qualifications will verify the true (real) and scientific notions as to their respective ideas.

- Further, the observer through his mind functions can verify the "statue" and the "Venus of Melos" aesthetic, intellectual and psychological functions as to the:

---

103 See Table 2 and 3.

- The observer verifies the underlying references of the same statue as to its beauty, its similarities with others, its historical and cultural context and so forth (aesthetic/artistic notions).

- Further, the observer describes his emotions related to the statue (ethics/activities notions).

Superlative-mind observer

In absolute terms, the superlative mind can by the definition proposed in this book, verify all the underlying knowledge required and truth related to the scientific, aesthetic/artistic, and ethics/activities notions and globally prove the immaterial ideas related to "Venus of Melos". In Plate 16 the superlative mind has the capacity to unquestionably verify the beauty of the statue of "Venus of Melos" as to its aesthetic/artistic qualities as well as all the complex notions related to the ethics/activities notions. The individual mind and even the advanced mind following the levels of mind defined in this book can only partly and subjectively describe these complex notions through his or her mind functions.

Scientific notions

Besides the true (real) situation notions, scientific notions encompass the entire Universe and all its entities. Scientific notions include the verification, observation, identification, experimental investigation and theoretical explanation of any natural phenomenon in the Universe.

Whenever such a notion is verified and proven mathematically or experimentally as true, then this notion is a scientific idea and becomes true knowledge. Until the verification and proof, the scientific notion remains a material stimuli, empirical idea, matrix, or thought pending for proof as not being yet a true scientific idea and remains as an opinion.

All three functions of the mind need to be operational for processing scientific notions; however, there is a prevalence of the intellectual functions.

As an example, the notion of water is a proven scientifically (chemical experimentation) entity and therefore forms a verified idea, at least for a knowledgeable chemist or scientist. Exact data of the properties of water are well known, proven and unquestionable[104] For all other persons the

---

[104] Water Properties   http://www.lsbu.ac.uk/water/data.html

water remains a visual and in general sensory generic notion perceived as a limpid, tasteless, drinkable, refreshing fluid and remains a material stimulus, or empirical idea or matrix or thought that corresponds to a true (real) notion limited to these qualities and only.

However, for the knowledgeable individual H_2O is a scientifically proven notion that constitutes an idea that fully describes the water and under experimental conditions this immaterial idea can reproduce the respective "water" as a material entity.

Aesthetic notions

In Plate 16 the observer of the "Venus of Melos", using the mind functions, will assess the beauty of the exhibit. The notion of beautiful is for the observer a thought that as a notion is subjective and not verified as a true idea; there will be a prevalence of the aesthetic functions related to the enchantment and fascination as to the magnificence.

The verification and upgrade of a stimuli, empirical idea, thought, or matrix to an idea depends on the level of mind. The verification "f" of the matrix of the notion "beautiful" referring to the statue can only be achieved as said earlier, by the superlative mind. The final verification of the notion referring to the beauty of the statue by the superlative mind classifies the aesthetic notion to an idea, which is by now true knowledge and not simply an opinion.

All other verifications "a, b, c, d, e" are processed by the individual, or the collective or the advanced mind without the ability to confirm the absolute "beauty" as can the superlative.

Notions related to ethics/activities

Ethics and compliant to ethics activities related notions differ from the scientific and the aesthetic/artistic notions. In the Plate 16 the presented statue provokes scientific, aesthetic/artistic and ethics/activities stimuli to the observer that can be variable including for example observation of the position of the statue in the Louvre, the lighting, the distance between the observer and the statue true (real) and scientific or the beauty (aesthetic) or the modesty of the gesture (ethics/activity).

Ethics/activities notions include those related to behavioral, psychological, moral, sociological, political, ideological, humanism faculties. Such notions are directly dependent in human societies to the civilization, educational, cultural, historical and ethnic practices and include responses to acquired civil, religious, societal, or moral impetuses or taboos. Ethics/activities notions include notions such as justice, virtuous, wise, and prudent as well as other relevant.

All three functions of the mind need to be operational for processing all the above notions; however, there will be a prevalence of the aesthetic functions to aesthetic/artistic notions, intellectual functions to scientific notions and psychological functions to ethics/activity notions.

NOOSPHERE AND NOOGENESIS

The human mind emerged during the evolution of species (Table 2 and 3).The mind as an information system is constantly evolving in Humans. The term noogenesis[105] in this book defines an evolutionary process in the Universe that leads to the development of the human mind and of the human societies.

The level of the mind in the last thirty thousand years shows an ascending trend that parallelizes the long-term trend of the evolution in general since the Big Bang (Plates 3 and 17). The variability of the genetic evolution and the changing environment determine the mind evolution and therefore its level is not synchronous in all parts of the Earth.

The levels of the mind include the individual, the collective, the advanced and the superlative based on the functions, the categories of notions processed and the performance. The individual mind (Tables 2 and 3) has evolved from the central nervous system body function to the early primitive mind before 350 to 250.000. This early mind level arbitrarily corresponds to the mind of a contemporary child of the age of 3 to 5 years in 2013 in an industrialized country.

The collective mind is the level of the human mind in 2013 extrapolated to the average mind of a modern western industrialized country's society. The advancement of the technology and of the civilization level of societies projects speculatively the future mind. An advanced mind in the future will reach the level that can process an ever-increasing number of scientific notions and verify their exactitude as ideas. It will also be capable to verify as well a limited number of aesthetic/artistic as well as ethics and activities notions.

---

[105] Definition adapted from the writings of Pierre Teilhard de Chardin who described first the emergence of intelligent forms as a noosphere in the evolution of humans. It also used in astrobiology concerning the emergence of forms of life capable of technology and so interstellar communication and travel.

The superlative mind will be the evolution of the collective future mind level and will verify and process all existing notions interacting with the totality of ideas (*logos*). The mind evolution from the individual to the collective and advanced mind can at any time retrograde to a lower level pending to external or genetic factors.

Plate 16 shows the circles that are representing the notions (immaterial ideas and *logos* (in dotted lines and grey color background) and material (in continuous lines in white background) concerning the levels of mind.

Logos

The *logos* is immaterial and includes all the ideas that correspond to verified, true and proven notions (real states/scientific, art and ethics/activities). *Logos* is thus the database of all immaterial notions that correspond equivalently to states, actions, events, snapshots and descriptions of the totality of all entities in the Universe. Every material entity that exists or existed is included as an idea representing its exact description and properties. The *logos* also includes the totality of the circles defined as "ideas" in Plate 15.

Among ideas are included:

- Those of the laws of Science

- All snapshots in the Universe(s)[106] since the Big Bang as being real states

- All the material mind notions (thoughts, matrices, empirical ideas, stimuli) that have been verified are tagged as verified ideas.

All the notions of possible errors, fantasies of the individual, collective and advanced mind levels as well as all possible numbers that are "nonsense" and do not consequently belong to an idea are not included in *logos*.

The *logos* is a purely immaterial entity comprising all the ideas. In turn, the ideas include all the existing states, entities and notions that are true. The *logos* as a state, has no space-time, is independent of any property, other state, condition, status or circumstance, and is free and independent from the material Universe. However, the ideas and in extension logos are compatible as a state with the material Universe.

---

[106] Universe(s) in the sense that outside the existing Universe where the Earth is the hypothesis that other Universes exist (multiverse theories)

The information notions of *logos* have the unique property that they are identical with verified notions as ideas of the mind; still, the mind is material ruled by the laws of Science and processing only material notions.

Interchangeability between immaterial and material

Every material entity has a dual expression, either as the object or as digital information that, when elaborated under suitable states and conditions, contains all the necessary description and details that will allow its exact reconstitution and reproduction.

In an earlier chapter[107], we elaborated the equivalence between immaterial and notions and real material entities with the illustration of the example of the iron rod. Another example of a typical interchangeability of an immaterial notion to a material is the genetic digital code of a virus. Plate 18 presents the experiment of reconstitution of the virus of influenza from a digital genetic code that is contained in eight genes to the reconstitution of the active living and infecting virus.

The genes of the influenza virus from an epidemic have been decrypted and the genetic code is a numeral encrypted in a four-letter nucleotide alphabet. This means that it is a pure abstract numeral, existing, that "is" and constitutes an idea. In special molecular genetic laboratories, it is possible to transform the digital information of the archived genes and resynthesize the original infective functional influenza virus.

This experimental model proves the interchangeability between immaterial notions and material entities. All living organisms and in general real entities are interchangeable to concrete numeral that exactly describes each at any space-time. The exact description is the snapshot: therefore, each snapshot corresponds to an exact and concrete numeral that by itself is an idea that exists and εἶναι. Certainly more the complexity of the material entity, more elaborated is the numeral abstract idea that is its equivalent. Furthermore, more complex and composite is the entity, even more sophisticated and complicated is its equivalent idea. Nowadays, advanced computer software in manufacturing can reproduce a concrete object that describes it in a digital numeral equivalence.

Mind involvement

The verification of all notions is a constant process of all levels of the mind. Despite that all levels of the mid are material, still their object of

---

[107] Equivalence paragraph in previous

action is the verification of the material notions, as discussed in the previous chapters.

Each level of the mind has categories of notions that it is habilitated to verify by privilege. Plate 19 shows the levels of the mind as to the three functions (aesthetic, intellectual, psychological) and as to their capacity for verification of the different categories of notions.

<u>Levels of mind</u>

Plate 20 shows the levels of the mind from the individual to the superlative, as to the increasing number of ideas (curve I). The superlative mind can process and verify all existing ideas (*logos*). In parallel (dotted curve II) is presented the curve of the number of unverified yet material notions that are decreasing as the level of superlative mind is attained. The superlative mind processes all three categories of notions (true state, scientific, aesthetic, ethics/activity) and can identify their correspondence to ideas.

The individual, collective and advanced mind processes also all categories of notions however can only prove their correspondence to ideas for a limited number of notions mainly true (real) states and scientific ones.

Curve I in the Plate 20 represents the totality of ideas (logos) that the superlative mind has the ability to verify. Curve II shows the gradually increasing number of true (real) state and scientific notions that the individual, collective and advanced mind that can verify and the minimal number of aesthetic and ethics/activities notions that are verifiable.

Plate 21 presents the number of verified ideas as to the material unverified ones. The angles φ, χ and ψ are measures of the ascending evolution of the levels of mind from the individual to the superlative. Theoretically, the superlative mind will be capable to verify and process all notions (angle $\psi 1$ with 90° angle) that will be the absolute number of existing ideas: the *logos*.

Plate 22 presents the parallel evolution tracks of the material Universe and the mind. This Plate illustrates the evolutionary ascending pathway from the Big Bang, through the main milestones of energy, elementary particles, atoms, molecules, living organisms and finally the human mind levels.

PHILOSOPHICAL MODELS

In the philosophy of the mind, idealism is in contrast to materialism, where the ultimate nature of reality is the physical substances. Idealism sometimes refers to a tradition in thought that represents things of a perfect form, as in aesthetics, ethics, morality, and value. In this way, it corresponds to the human perfect being or state.

Idealism is a philosophical movement in Western thought, and delineates philosophical positions with various tendencies and implications in politics and ethics; for instance, philosophical idealism is associated with Plato and the school of Platonism. From the other side, materialism holds that the only thing that exists is matter; that all things are composed of material and all actions, states and functions (including the mind) are the result of material interactions.

In this book, we propose the "information model" as a separate philosophical model based on the material information system of the human mind having the ability to process notions and verify them as to their conformity to true and proven respective ideas.

The basis of this model is the proposed earlier in this book Psychobiology described by the four operational ranks: functions, programs, methods and tools. There is separation of the material appurtenance of the mind in the human body from the immaterial information; the mind operating as a purely material information system.

Plate 14 depicts the description of the information model. A novelty of the proposed information model is the maintenance of the Platonic ideas however keeping the information notions material and external to the human mind.

The link between the immaterial ideas and the material notions is only through the verification in a bottom to top approach with the gradual upgrade of the human mind to the superlative level. Another link is the bridging of the ideas between the three states: the *logos*, the *beyond*[108] and the human mind. The bridge is from one hand the mind, material in nature, and from the other the immaterial notions; both interact through

---

[108] In the next chapters the *beyond* is defined as "a state that exists (εἶναι) and is characterized as free of any property or variable or whatever dependence such as space-time, energy, fields and of any other attribute or dimension or submission to the laws of Science except the property to bounce from virtual real particles." The term has been transposed from Plato's ΕΠΕΚΕΙΝΑ ΤΗΣ ΟΥΣΙΑΣ. Politeia 509 B which is defined within the context of the "Αγαθόν, good" while in this book the term is related to both Physics and Philosophy. This is defined in this book by the *beyond* situation which is the absolute "good" compatible with ideas and the logos from which all Universes are automatically cerated. .

the verified ideas of the superlative mind. The ideas in this sense remain an entity of the *logos*.

The use of the proposed information-philosophical model based on the superlative mind, the *beyond* and the *logos* unifies the idealism and materialism: the immaterial notions are accessible to the material human mind through the superlative mind and even further, the individual, collective and advanced mind levels share this accessibility in a limited degree.

How now does the Universe evolution link to the *logos*? The bridge again is the human mind through which the mind levels, from top to bottom, process the immaterial notions and through which proceed to active indeterministic actions aiming to active purposes. Under the rule of the laws of Science and with the governance of determinism, indeterminism and randomness, this evolutionary process translates to the ascending trend of the Universe observed on Earth.

Ascending pathway

The Ancient Greeks had defined as the utter objective of the world the tecmor[109] meaning objective, target, and goal or even fixed mark or boundary, end or purpose. Alcman describes that starting from an amorphous mass, the guidelines or the contriver (poros) lead to the tecmor under the tunneling of Thetis that is the operator. This simulation gives the dynamics of an evolutionary process starting from an unformed entity and aiming to a purpose or target. In this book, by extrapolating Alcman, the amorphous starting matter represents the primordial soup, the poros the ruling of the laws of Science, Thetis the developmental process of the Universe and tecmor the superlative mind.

Plate 23 on the left depicts the principle events that mark the evolution in the Universe. On the left are designed route maps representing series of snapshots (monoclonal or polyclonal) leading to the appearance of the mind. As discussed in previous chapters, since 10^{-34} seconds after the Big Bang to date, a continuous sequential chain of events registers snapshots that follow the pathway potentially leading to a theoretical superlative

---

[109] In one fragmentary hymn (by the seventh century Spartan poet, Alcman), Thetis appears as a demiurge, beginning her creation with poros (πόρος) "path, track" and tekmor (τέκμωρ) "marker, end-post". Empedocles, later described the Physiology of the sensory transmission by extrapolating the notions of πόρος to οπή (hole) and πόρος to απορροή (flux).

mind. Some of the snapshot series follow route-maps that lead to the human mind while others take different evolutionary routes. Some of these routes are leading to extinction others continue until now probably leading to other targets that are unrelated to the evolution towards a mind.

The basis of Plate 23 is the prerequisite for achievement at each echelon of all the conditions succeeded in the previous points. This successful pathway is the final echelon "ν" and the route-map followed the echelons "1, 2, 3, 4, 5, 6, 7, 8, 9, μ..., ν-1 and ν".

Again, Plate 23 presents the possibility of more than one superlative mind. It is also demonstrating the possibility of pathways failing to lead to the superlative mind; in this case following a failed or discarded or aborted or extinct pathway (with echelons ... μ', ν'-1 and ν' instead of the successful 1, 2, 3, 4, 5, 6, 7, 8, 9, μ..., ν-1 and ν of this Plate.

The route-map that led to *H. sapiens* and the mind of the contemporary inhabitants of the Earth do not necessarily lead to a higher level of mind or to the superlative and might be either dead ends or extinctive. Again, Plate 23 shows that there is a probability that the civilization on Earth is a failure or a dead-end and does not lead to a higher mind level.

Possibly, other route-maps might exist, belonging to different evolutionary series of snapshot with more successful pathways that lead to a superlative mind. Such pathways already exist or have existed in the past or will exist in the future somewhere in the Universe outside the Earth.

The mind and the human societies on Earth empirically and based on the historical evolution follow an ascending trend. This is an objective proven observation: the *H. sapiens* of the caves before thirty-five centuries and the contemporary sophisticated and highly educated and civilized societies are specifically an advance. The route-map pathway to a more advanced mind or even a superlative mind is potentially possible however uncertain at the rhythm by which history is evolving on Earth.

Restriction in the evolutionary process

The possible evolutionary pathways toward the emergence of a human mind are inflexible and restricted because of the interplay of the five variables, the laws of Science and the interacting environmental changing conditions.

In a way, the pathways shown in Plate 23 allow within the restrictions a certain degree of freedom, unpredictability and flexibility based on occasional alternation of governance from determinism to randomness. On the contrary, the five variables, the governance and the environmental and historical circumstances in the Universe from the Big Bang to the mind globally restrict the flexibility of the evolutionary process. The restriction is more evident because of the interplay of the five variables on Earth where the increasing number of biochemical molecules in living organisms renders the evolution highly complex, competitive and variable. All five variables interplay in the rapidly evolving Universe after the formation of the celestial bodies and especially after the appearance of living organisms limit determinism and reduce predictability.

There is no doubt and it is self-evident that a route-map pathway of the Universe evolution is the one that led to Humans on Earth. However, this does not mean that this was the only route-map and that a considerable number of other routes have not led to different evolutions. It is also possible that if exactly the same route-map that led to the Humans on Earth led also to the appearance of *H. sapiens* in other exoplanets. The estimate is that some 10^{13} stars are present in the Universe, therefore the probability of Humans in other exoplanets is not negligible, knowing that as a rule stars orbit planets.

The flexibility was mostly limited at the early phases of the evolution to the formation of the celestial bodies. Later, during the biological evolution until Humans, the genetic variations allow a degree of freedom and variance. However, each period in the evolution constitutes a different interplay of the five variables and the time sequences of the snapshot series. The governance of randomness is omnipresent on the level of quanta mechanics and of the interplay of the five variables, however globally insignificant because of mutual neutralization in classical Physics.

The history of the Universe evolution including the Earth has crossed concrete echelons: elementary particles, energy, atoms, molecules, nucleotides, RNA/DNA, universal common ancestry, bacteria, animalia, *H. sapiens* and the human mind.

The early history until to the formation of Earth consisted of a natural sequence of physical events that evolved with a limited flexibility for alternative developments that ended with a picture that is not different from what observed today in the Universe, the celestial bodies and on Earth. Later, the evolution on Earth with the creation of the biosphere allowed an increased flexibility as to the evolutionary pathway that leads to Humans and their mind.

Statistically the probabilistic calculation that similar Human-like communities exist in other exoplanets is high based on recent observations and discoveries. The probability is imperatively increased because of the high number of stars in each galaxy and the number of galaxies in the Universe.

Noocentric evolution

The modulation of the tetractides of elements in the evolutionary process of the Universe, achieved the ascending trend and the human mind appearance. The pathway followed allowed, by modulating quanta, elementary particles, atoms, molecules and genetic codes through the expression of compatibilities and restrictions, the emergence of the contemporary Universe that we observe. The noocentric evolution is the pathway followed in the evolution that led to the mind.

The pathways in the evolution are potentially multiple complying with the restrictions of the five variables and the governances of randomness or determinism (indeterminism appearing only after the human mind).

At the end, the historical observation, traces the pathway that led to the noocentric evolution leading to the mind as observed on Earth today and to the speculation that eventually a superlative mind can be reached in the Universe.

Trend to superlative mind

Plate 24 raises the question if there is a possibility in the Universe to reach to more than one superlative mind. Theoretically - based on the definition of the superlative mind as an evolutionary consequence that can verify and process and all notions if they correspond to ideas - the possibility of coexistence of more than one superlative mind is valid.

In this case, we raise the following question: can more than one superlative mind coexists and not be competing?

The answer to this ethical question is that certainly more than one superlative mind can exist; however, each processes different in each case notions to verified ideas depending on the specific environmental conditions where operating. Therefore, the superlative mind has no other possibility than to perform the ideally "good", however in each separate condition. Plate 24 shows the plurality of the different levels of mind that potentially tend to one or more superlative mind.

"Good" and superlative mind

The assumption of "good" refers to the superlative mind that acts based on verified notions that are real (εἶναι), proven by science and indisputably beautiful, justice, virtuous, wise or prudent or of any other qualification within.

In this book, the definition of "good" is a combination covering beauty, fineness, excellence, goodness and unity. The basis of this definition is Plato's of τὸ ἀγαθόν and τὸ καλόν as well as Aristotle's of εὖ and of αὐτὸ τὸ ἓν τὸ ἀγαθὸν αὐτὸ εἶναι.

Due to the broad definition of "good" here, the notion of "evil" is not used as the opposite and instead the notion of "non-good" is preferred as such. "Good" in this book is a tecmor of creation and evolution and includes the Philosophical as well as the Physics meaning of the notion.

<u>The superlative mind pathway</u>

In all cases not leading to a superlative mind, the pathway followed will at some point fail to respond and will be apt to actions that can be destructive and "not good" therefore incompatible to the definition of superlative mind.

In the unique case where all the conditions in two or more pathways leading to a superlative mind the question raised is: are these superlative minds identical replicas? The answer is again positive since the notion of "good" as an idea is unique and therefore all the potential superlative minds will be replicas. However, each develops through a potentially different pathway, which depends on the states occurring in each evolutionary process followed. These superlative minds will function independently pending on the environmental conditions and the governance but always inside the ideally "good".

Plate 24 shows the multitude of potential mind levels. A further question is if an advanced mind can perform actions that are "non-good" that can be destructive. The answer again is positive however this mind will never reach the superlative level.

UNIVERSES

Recently several theories and concepts tend to show that the Universe, where the Earth is included, is not the only such but many other Universes can sprout from existing ones based on the inflationary Universe model that is in itself an extension of the Big Bang theory.

The evolution on Earth with the development of the human mind is compatible with similar potential evolutions in other Universes. Even inside the existing Universe where the Earth is, the possibility of evolutions on other exoplanets that orbit a star other than Earth's sun, theoretically can lead to the development of similar to the human mind as observed on Earth. The restrictions based on the laws of Science, the five variables and the environmental conditions of the evolving snapshots are applicable like on Earth and elsewhere; however, diversity within these limits is possible.

The end-point of the evolution in noocentric terms is the superlative mind. This endpoint, as tecmor, therefore is the same for other exoplanets and even for other Universes. The evolution wherever taking place aims to a superlative mind; however, usually it leads to a failure or to an abortive mind, faraway from reaching the level of the superlative mind.

One other theoretical subject discussed recently by Steve Hawking[110] is the possibility of a Universe creation in a laboratory. We name such a Universe as "techno-verse". Following the reasoning in the previous chapters on the mind, this is plausible for an advanced and certainly for a superlative mind in a future human high-technology society. In such a case, Humans become creators of Universes.

IMMORTALITY AND ETERNITY

The digital code of the DNA at the zero time of conception corresponds to a concrete human at that instance (Plates 8 to 10). For example, the genetic information contained in the ovum of a mammalian, at the fusion moment between the paternal and maternal cell, contains all the information required. The chain of four radicals of nucleotides inscribes this DNA (or RNA) code, which is an immaterial idea. At the same time, based on the equivalence described in an earlier chapter, the same numeral four-letter nucleotide-code is a real material entity that describes the concrete human ovum for which it stands for.

This code becomes complex and splits to numerous sub-codes that are all changing following the development of the conceived human from conception, through birth and death. The environmental conditions existing at each instance of the development of the conceived concrete human change constantly the code and the genetic information in the newly produced cells of the embryo and later the newborn human being.

The numeral digital code at the instance of conception is the verified corresponding immaterial idea. The future human being becomes

---

[110] The Grand Design, Stephen Hawking and Leonard Mlodinow, Published by Bantam Books, 2010.

interchangeable and is exactly reproducible under identical environmental and experimental conditions.

At the same time the digital code as an idea is immaterial and free of space-time dependence; therefore, this notion is eternal. Furthermore, there is expression of this same idea as a numeral digital code, to the respective concrete Humans, whenever suitable environmental or experimental conditions occur.

As with Humans, any biological living organism will have the same equivalence between the immaterial idea of its genetic information and the corresponding to it real biological being.

Therefore, all viable genetic codes as ideas that correspond to reciprocal human or other living organisms are ideas included in the *logos*. As such, these ideas are compatible with the *beyond*, free of all properties including space-time.

Option on future evolution

On Earth in 2013, a remarkable evolution of human societies is occurring and the collective mind is heading to the level of the advanced mind. This rapid evolution seems to be a unique phenomenon based on the discoveries in science and the improvement in civilization standards.

There are diverse evolutionary pathways aiming to reach the level of a superlative mind (Plate 23). Among these pathways, by observing the evolution of the human societies on Earth, there is an evident ascending trend. This is in particular manifest if we observe the history in the past four thousand years since the early Hittite and the Sumerian civilizations. Even more, in the past few decades, there is a strong acceleration of this trend based on the technological and social advances in many societies.

However, despite the historical evidence of the ascending track in evolution on Earth, it is impossible to speculate on the perspectives on the evolution of human societies. Therefore, it is impossible to affirm that the level of human mind will ever reach the superlative mind. This does not exclude that there exists a superlative mind, reached under similar to the Earth conditions in an exoplanet in the actual Universe or in other potential Universes.

PART 4 GENERAL THEORY OF EVOLUTION

The Universe where the Earth subsists is an entity that includes all matter and energy including the galaxies and the galactic intermediary space. In more detail, the content contains approximately 4.6% atoms, 23% of dark matter that does not emit or absorb light and 72% dark energy responsible for the acceleration of the Universe.

Ongoing research indicates that the dark matter might be due to positrons in the Universe. Recently, the CERN[111] announced results that are based on some 25×10^9 recorded events, including 4×10^5 positrons with energies between 0.5 and 350 GeV[112] , recorded over a year and a half. This represents the largest collection of antimatter particles recorded in space. These results are consistent with the positrons originating from the annihilation of dark matter particles in space, but not yet sufficiently conclusive to rule out other explanations.

Further, according to recent experimental data, multiple Universes bourgeoned as ramifications from a primary one. For the simplification of discussion, we will focus on the Universe where today the Earth is, this not meaning axiomatically that this is the primary, initial or unique one. We refer to a primary scalar field that we propose to be the initial causative creative start for every Universe evolution.

THE *BEYOND* AND IMMATERIAL NOTIONS

The definition of the *beyond* given in the previous chapters, is that it is a state that exists (εἶναι) and is characterized as free of any property or variable or whatever dependence such as space-time, energy, fields and of any other attribute or dimension or submission to the laws of Science except the property to bounce from virtual real particles. It coexists independently and unrelated to ideas and *logos* (that comprises the totality of all ideas). As to the laws of Science, they are valid in the *beyond* as

---

[111] CERN Centre Européen des Etudes Nucléaires.
http://web.mit.edu/newsoffice/2013/cern-announces-measurement-of-antimatter-excess-in-space.html

[112] gigaelectronvolts, a unit of energy equal to 1×10^9 electron volts

immaterial notions (ideas), however their applicability is nil since there is no object for their appliance.

The *beyond* has no time dimension and therefore, has no conventional chronological measurement as in the Universe where time starts just at the Big Bang.

As reminder, we refer to the ideas as immaterial abstract immutable information notions proven by science and indisputably beautiful, justice, virtuous, wise or prudent as verified by the superlative mind (that acts only in "good" actions).

VIRTUAL PARTICLES AND THE *BEYOND*

We will define the bouncing (or creation) of a virtual particle in the Universe firstly and secondly in the *beyond*.

In the Universe the uncertainty principle implies that particles, defined as virtual[113], can come into existence for short periods even when there is not enough energy to create them and in effect, their creation is due to uncertainties in energy. Based on the uncertainty principle there is a violation of the energy conservation when for example a pair of an electron and of a positron bounce for a very short time of the order of 10^{-20} seconds. The particle and its antimatter particles will briefly "borrow" the energy required for their creation, and then, a short time later, pay the "debt" back and annihilate.

The virtual particles in the Universe are part of the quanta mechanics that describe fields of the basic force interactions. The degree of uncertainty of each is inversely proportional to time duration (for energy) or to position span (for momentum). Virtual particles exhibit some of the phenomena that real particles do, such as obedience to the conservation laws.

The zero-point energy is the lowest possible energy that a quantum mechanical physical system may have; it is the energy of its ground state. All quanta mechanical systems undergo fluctuations even in their ground state and have zero-point energy associated, which is a consequence of their wave-like nature. The uncertainty principle requires every physical system to have a zero-point energy greater than the minimum of its classical potential well, even at absolute zero temperature.

---

[113] The Edges of Science: Crossing the Boundary from Physics to Metaphysics. Richard Morris. Prentice Hall 1990.

Vacuum energy is the zero-point energy of all the fields in space, which in the Standard Model includes the electromagnetic field, other gauge fields, fermion fields, and the Higgs field. It is the energy of the vacuum, which in quantum field theory is not completely empty space but as the ground state of the fields. In cosmology, the vacuum energy is one possible explanation for the cosmological constant. A related term is zero-point field, which is the lowest energy state of a particular field.

PRE-BIG BANG CONDITIONS

In the *beyond*, we postulate that the virtual particles can interchange and bounce to real, as it occurs in the Universe. This is described in the next paragraphs in five steps:

First step: the pre-Big Bang Universe creation

Second step: narrowing to the observance evidence of the Universe where the earth lies

Third step: from virtual to real particle

Fourth step: the bouncing event

Fifth step: from real particle to a Universe

Sixth step: from real particle to scalar field

Seventh step: evolutionary process

The new hypothesis launched in this book is founded and is compatible with the following four principles:

1. Casimir effect[114]

---

[114] In quantum field theory, the Casimir effect and the Casimir-Polder force are physical forces arising from a quantized field. The typical example is of two uncharged metallic plates in a vacuum, placed a few micrometers apart, without any external electromagnetic field. In a classical description, the lack of an external field also means that there is no field between the plates, and no force would be measured between them. When this field is instead studied using quantum electrodynamics, it is seen that the plates do affect the virtual photons which constitute the field, and generate a net force—either an attraction or a repulsion depending on the specific arrangement of the two plates.

2. Heisenberg uncertainty principle

3. Dirac's delta function[115]

4. Dirac's superposition[116]

The principles 3, 4 and 5 are discussed in step 2, 3, 4 and 5.

Although the Casimir effect can be expressed in terms of virtual particles interacting with the objects, it is best described and more easily calculated in terms of the zero-point energy of a quantized field in the intervening space between the objects. This force has been measured, and is a striking example of an effect purely due to second quantization. However, the treatment of boundary conditions in these calculations has led to some controversy. In fact "Casimir's original goal was to compute the van der Waals force between polarizable molecules" of the metallic plates. Thus it can be interpreted without any reference to the zero-point energy (vacuum energy) or virtual particles of quantum fields.

Dutch physicists Hendrik B. G. Casimir and Dirk Polder proposed the existence of the force and formulated an experiment to detect it in 1948 while participating in research at Philips Research Labs. The classic form of the experiment, described above, successfully demonstrated the force to within 15% of the value predicted by the theory.

Because the strength of the force falls off rapidly with distance, it is only measurable when the distance between the objects is extremely small. On a sub-micrometer scale, this force becomes so strong that it becomes the dominant force between uncharged conductors. In fact, at separations of 10 nm (about 100 times the typical size of an atom)the Casimir effect produces the equivalent of 1 atmosphere of pressure (101.3 kPa), the precise value depending on surface geometry and other factors.

In modern theoretical physics, the Casimir effect plays an important role in the chiral bag model of the nucleon; and in applied physics, it is significant in some aspects of emerging microtechnologies and nanotechnologies.

http://www.andersoninstitute.com/casimir-effect.html

[115] The Dirac delta can be loosely thought of as a function on the real line which is zero everywhere except at the origin, where it is infinite, $\delta(x) = \begin{cases} +\infty, & x = 0 \\ 0, & x \neq 0 \end{cases}$ and which is also constrained to satisfy the identity $\int_{-\infty}^{\infty} \delta(x)\,dx = 1.$ This is merely a heuristic characterization. The Dirac delta is not a function in the traditional sense as no function defined on the real numbers has these properties. The Dirac delta function can be rigorously defined either as a distribution or as a measure. Dirac, Paul (1958), Principles of quantum mechanics (4th ed.), Oxford at the Clarendon Press, ISBN 978-0-19-852011-5.

[116] At this point the general principle of Dirac with the use of the quantum superposition of quantum mechanics comes-in describing "how a physical system such as a real elementary particle in this case, exists partly in its entire particular, theoretically possible states simultaneously; but, when measured or observed, it gives a result corresponding to only one of the possible configurations"

CASIMIR EFFECT AND ZERO-POINT ENERGY

We have described the *beyond* in earlier chapters. Zero-point energy[117] is an existing proven state: its main difference with the *beyond* is that the zero-point energy disposes quanta fields and is defined within the Universe. The *beyond* is by definition free of all fields, functions, variables and other properties, excepting the function of bouncing from a virtual to a real particle).

In the context of the beyond and the Casimir effect the vacuum is defined as follows:

"The vacuum[118] has, implicitly, all of the properties that a particle may have: spin, or polarization in the case of light, energy, and so on. On average, all of these properties cancel out: the vacuum is empty in this sense. One important exception is the vacuum energy or the vacuum expectation value of the energy. The quantization of a simple harmonic oscillator states that the lowest possible energy or zero-point energy that such an oscillator may have is $E = 1/2\hbar\omega$ ..".

"In the situation of the Casimir effect and the equations and mathematics describing it[119], the force in the area between the two plates is described:

The Casimir force per unit area F_c / A for idealized, perfectly conducting plates with vacuum between them is

$F_c / A = -d(E)/da \; A = -\hbar C\pi^2 = 240a^4$ where $\hbar$ is the reduced Planck constant, c is the speed of light, a is the distance between the two plates.

The force is negative, indicating that the force is attractive: by moving the two plates closer together, the energy is lowered. The presence of $\hbar$ shows that the Casimir force per unit area F_c/A is very small, and that furthermore, the force is inherently of quantum-mechanical origin".

The concept of zero-point energy module is extrapolated to the state of the Casimir effect[120] and further in this book where we have introduced

---

[117] Albert Einstein and Otto Stern developed the concept of zero - point energy in Germany in 1913, using a formula developed by Max Planck in 1900.

[118] http://www.andersoninstitute.com/casimir-effect.html

[119] http://phys.lsu.edu/~jdowling/PHYS4112/Dowling89d.pdf
http://milesmathis.com/casimir.html

[120] The causes of the Casimir effect are described by quantum field theory, which states that all of the various fundamental fields, such as the electromagnetic field, must be quantized at each and every point in space. In a simplified view, a "field" in physics may be envisioned as if space were filled with interconnected vibrating balls and springs, and the strength of the field can be visualized as the displacement of a ball from its rest position.

the notion of the *beyond*. In this context the *beyond* is the absolute "vacuum" and at the same time it is the absolute Casimir effect if the plates are introduced in an absolute vacuum.

As to the virtual particles these are classically considered as particles which flash into and out of existence spontaneously. They are allowed to "borrow" rest energy via the uncertainty principle $\Delta E \Delta t \geq \frac{1}{2} \hbar,$ but only for a short time Δt. The Casimir effect discussed below is an attraction between two plates in a vacuum caused by virtual particles.

First step: the pre-Big Bang Universe creation

By extrapolating to the *beyond* state the Area is zero as is also the physical existence of the two plates and therefore the distance between them. However the expectation value of the energy (F) remains and in this case is infinite (ἄπειρον ∞). Assuming that the *beyond* state the distance and the area between the plates in Casimir effect is and the total force (F_c/A) is infinite.

The *beyond* state as in the Casimir effect will create infinite virtual particles always at a zero time and space dimension.

Vibrations in this field propagate and are governed by the appropriate wave equation for the particular field in question. The second quantization of quantum field theory requires that each such ball-spring combination be quantized, that is, that the strength of the field be quantized at each point in space. Canonically, the field at each point in space is a simple harmonic oscillator, and its quantization places a quantum harmonic oscillator at each point. Excitations of the field correspond to the elementary particles of particle physics. However, even the vacuum has a vastly complex structure, all calculations of quantum field theory must be made in relation to this model of the vacuum.

The vacuum has, implicitly, all of the properties that a particle may have: spin, or polarization in the case of light, energy, and so on. On average, all of these properties cancel out: the vacuum is, after all, "empty" in this sense. One important exception is the vacuum energy or the vacuum expectation value of the energy. Summing over all possible oscillators at all points in space gives an infinite quantity. To remove this infinity, one may argue that only differences in energy are physically measurable; this argument is the underpinning of the theory of renormalization. In all practical calculations, this is how the infinity is always handled. In a deeper sense, however, renormalization is unsatisfying, and the removal of this infinity presents a challenge in the search for a Theory of Everything. Currently there is no compelling explanation for how this infinity should be treated as essentially zero; a non-zero value is essentially the cosmological constant and any large value causes trouble in cosmology. http://www.andersoninstitute.com/casimir-effect.html

"Virtual particles[121] do not necessarily carry the same mass as the corresponding real particle, although they always conserve energy and momentum. The longer the virtual particle exists, the closer its characteristics come to those of ordinary particles. Virtual particles may be thought of as arising due to the time-energy Heisenberg's uncertainty principle. They are important in the physics of many processes, including particle scattering and Casimir effect. In quantum field theory, even classical forces — such as the electromagnetic repulsion or attraction between two charges — can be thought of as due to the exchange of many virtual photons between the charges".

The term is somewhat loose and vaguely defined, in that it refers to the view that the world is made up of "real particles": it is not; rather, "real particles" are better understood to be excitations of the underlying quantum fields. Virtual particles are also excitations of the underlying fields, but are temporary in the sense that they appear in calculations of interactions, but never as asymptotic states or indices to the scattering matrix. As such the accuracy and use of virtual particles in calculations is firmly established[122].

Each virtual particle created will still remain at a zero time and space dimension till it will bounce to a real particle. At this point for each virtual particle time, space and all fields, functions, variables and other properties will be created for the newly created real particle. The conditions will be therefore created that each single bounced real particle will create its own Universe. The newly created particles will be completely independent from all other particles and there will be no link between them or between them and the *beyond* state.

The required energy for the bouncing is in this case independent from the F_c/A of the Casimir effect equation, however as in the case of the Universe this energy must be momentarily borrowed. In this case of the newly created real particle (and at the same time new Universe) the return of the energy cannot be rendered back to the *beyond* state, however it will follow its evolution in the newly created individual time-space.

Plate 25 presents the two states: up the zero-point energy and down the *beyond* state. Both the zero point energy and the *beyond* have a common property that they can bounce automatically a virtual particle to a real

---

[121] https://en.wikipedia.org/wiki/Virtual_particle

[122] .. but their "reality" or existence is a question of philosophy rather than science.. In this book since the virtual particles are scientifically and experimentally provable then they exist and therefore are real.

material particle. In the *beyond* where there is no start as is the case in the Universe, the bouncing is potentially infinite occurring statically while it remains *per se* unmovable.

In this plate in the *beyond* state the respective virtual particle, as an idea, remains immaterial: we define the potential virtual particle bouncing to real as the iconic virtual particle. The iconic virtual particle we propose is the virtual particle that remains as an idea, immaterial and has not acquired the properties of the virtual particles[123] and by extension has not yet bounced to real as defined in Physics.

Narrowing the observation of the Casimir effect (F_c/A) to the specific particle that corresponds to the real particle that was the start of the Universe where the Earth lies, we link the *beyond* with the pre-Big Bang zero time-space condition. Each of the other virtual particles will or will not bounce to a real pending on each particle's properties. But even if bounced to a real again it can or cannot evolve to a Big Bang situation as is the case of the Earth's Universe.

Second step: narrowing to the observance evidence of the Universe where the earth lies

The observance is only realized *a posteriori* and only in the created Universe and solely by the human mind. In the state of the *beyond*, while there are no fields, functions, variables and other properties (excepting the function of bouncing from a virtual to a real particle) however the laws of Science are compatible. All particles virtual or real provably exist both as ideas as well as defined virtual or real particles and therefore are part of the *logos*.

In the concrete case of the Universe where the Earth lies, from a virtual bounced to real particle material entity is created out of immaterial (*beyond* state). The observance object is the Universe where the earth is, the evolutionary link between this particle and the Big Bang is proposed in a stochastic model presented in the next chapter.

---

[123] Some of the field interactions which may be seen in virtual particles in the Universe are: the Coulomb force, the magnetic field between magnetic dipoles, the so-called near field of radio antennas, the strong nuclear force between quarks the weak nuclear force - it is the result of exchange by virtual W bosons, the spontaneous emission of a photon during the decay of an excited atom or excited nucleus; such a decay is prohibited by ordinary quantum mechanics and requires the quantization of the electromagnetic field for its explanation, the Casimir effect, where the ground state of the quantized electromagnetic field causes attraction between a pair of electrically neutral metal plates, the van der Waals force, which is partly owing to the Casimir effect between two atoms, the vacuum polarization, the Lamb shift of positions of atomic levels, the Hawking radiation

The Dirac[124] delta (δ) function applies at this state between the virtual and real particle when all the functions are zero except one the integral of "one" over the entire real line. The zero exception is the point mass or point charge that constitutes an integral of "one" over the entire real line. The function of the bouncing of the virtual to the real particle to real, is considered in this book as an infinitely high, infinitely thin spike at the origin, with total area one under the spike.

Third step: from virtual to real particle

The hypothesis proposed is that the *beyond* automatically and spontaneously at the zero time bounces a virtual particle to a compatible real. This is assumed by extrapolating the Casimir effect where the area (A) and the distance between the plates of the original experiment are set at a zero value. This assumption, as described earlier, will render in the equation the F at a value of infinite.

The hypothesis proposed is that the *beyond* automatically and spontaneously at the zero time bounces a virtual particle to a compatible real. This is assumed by extrapolating the Casimir effect where the area (A) and the distance between the plates of the original experiment are set at a zero value. This assumption, as described earlier, will render in the equation the F at a value of infinite.

The events that link the *beyond* state to the real particle that has created the observed Universe where the earth lies is as follows:

Beyond>logos>ideas>idea   describing   a   concrete   iconic   virtual particle>iconic   virtual   particle>virtual   particle>Casimir   effect conditions>real particle>new Universe>Big Bang>evolutionary process as described in Physics

The virtual particle is an idea that exactly, interchangeably and absolutely describes the respective real particle and is included in the database of the *logos*, therefore compatible with the *beyond*. We assume that the *logos*, which is compatible with the *beyond*, disposes the totality of the real particle ideas

---

124 The Dirac delta function, introduced by Paul Dirac, can be informally thought of as a function $\delta(x)$ that has the value of infinity for $x = 0$, the value zero elsewhere, and a total integral of one. The graph of the delta function can be thought of as following the whole x-axis and the positive y-axis, is a generalized function on the real number line that is zero everywhere except at zero, with an integral of one over the entire real line. The delta function is sometimes thought of as an infinitely high, infinitely thin spike at the origin, with total area one under the spike, and physically represents an idealized point mass or point charge.

constituting a registry. This registry of the iconic virtual particles is ruled by the laws of Science and every single idea within it is scientifically proven and real.

The time in the *beyond* does not exist and is represented in Plate 25 by a T (time) value of zero, while in the Universe where the state of the zero point energy is, the relative time is valid (starting from the creation of the Universe at the Big Bang). The T (time) remains zero during the interchangeability from virtual to real in both the *beyond* and the zero-point energy states. Plate 26 describes the events that link the *beyond* state to the evolution up to the human mind on Earth.

Fourth step: the bouncing event

This bouncing initiates the start of time during which the real particle "borrows" energy for an extremely brief period ($<T$). Starting from the instance when the bouncing occurred, the real particle will acquire all the material properties of particles, including the variable of time (T short).

The iconic virtual particle as well as the virtual particle exhibits[125] the Casimir effect. On the contrary, in the zero-point state, the virtual particles bouncing for a short period are couples of particle-antiparticle that rapidly annihilate and exhibit the known properties of virtual particles as described in Physics. The iconic virtual particle is identical to the virtual particle: both being provably and experimentally defined by Physics. The iconic virtual particle remains an idea however the virtual particle is when it can be manifested in a Casimir effect experiment. Furthermore, iconic as well as the virtual particle can only be observed *a posteriori* whenever it is actualized and conceived by the human mind.

Specifically in the zero-point energy state, the real particles are annihilated which is not the case in the *beyond*. Every virtual particle bouncing from *beyond* to real creates a new state, which corresponds to all the variables and properties of a Universe. In a similar condition from the *beyond* state where a couple of positive and negative particles are bounced each one will create a separate Universe and therefore there will not be annihilation:

---

[125] The couple of virtual and real particle exist as interchangeable two natures of the same notion: immaterial as a virtual particle and material as its corresponding real particle. The now real particle while it lasts, disposes space dimensions, and remains material; it follows the laws of Science (laws of physics and of quantum mechanics). The return of the borrowed energy back to the spaceless and timeless beyond is irrelevant, because in the material Universe the beyond is not an existing receptor destination.

each bounced to real particle is independent from its bounced antiparticle and each exists in a separate Universe.

The iconic virtual particle as well as the real particle, are (εἶναι) in a similar manner as in the zero-point state in the Universe where a virtual particle can potentially bounce to a real one (see Plate 26). These are defined in this book as converses.

The converses possess all the fields, functions, variables and other properties, which describe the material Universe and are compliant to the laws of Physics. However not all will evolve further to a Big Bang. The converses that potentially will have all the fields, functions, variables and other properties as was the case with the actual Universe where the earth lies will evolve possibly even to the appearance of a superlative mind. Others will evolve to earlier stages of a Universe: the evolution stage is made by comparing the proven, real and observed evolution of our own Universe.

The bounced particles as converses can each either remain as a particle creating its own Universe or even it can initiate a process towards an evolutionary new Universe that is abortive as to the endpoint reaching any evolutionary point prior to the appearance of a human mind.

Since time does not exist in the state of the *beyond*, the bouncing of the virtual particles is infinite based on the Casimir equation where the F (Casimir force) is infinite. The virtual particles created are real, omnipresent and infinite in number. The time dimension renders the infinity since at the *beyond* state $T=0$ and for each newly created virtual particle in the Casimir effect conditions the time non zero.

In both the zero-point energy Universe and the *beyond* state, the virtual and the respective real are interchangeable linked by a notion that is abstract information and defined as an idea (in this case as the iconic virtual particle). The immaterial idea of an iconic virtual particle exists and the proof of its existence is its interchangeability to the corresponding real. For example, the idea describing a concrete electron[126] (e−) that it can be either virtual or real, is the fact of its scientifically provable, real, true,

---

[126] With distinct properties (among others that it is a fermion and more precisely a lepton, having an antiparticle, the positron, with an electrical charge (-1e)of $1,6\times10^{-19}$ Coulomb, a mass of $9,1\times10^{-31}$ g, and a spin of $\frac{1}{2}$

and verified existence (εἶναι). The same is for its antiparticle, the positron[127].

In the example of an electron or a positron after bouncing from the virtual to real, they possess concrete fields, functions, variables and other properties, scientifically and experimentally proven corresponding to a respective idea. An "α" particle that will finally end up as evolving to a successful Universe like the one where the Earth lies is defined as a primary "α" particle and it will evolve towards an "α" primary scalar field, an "α" Big Bang, an "α" Universe explosion that can follow the evolutionary process towards the human mind or even the superlative mind or it can stop at any intermediate point.

The mind can conceive the *beyond* as a state or situation only after the existing conditions in the Universe are evaluated, defined and proven. The exception property for the creation from a virtual to a real particle takes into account that the *beyond* is compatible with the immaterial *logos* (that includes all ideas) as well as the zero-point energy state in the material Universe.

This exception is based on Dirac's General Principle. In this book the condition of the existence of the *beyond* corresponds to the definition of Dirac for the "state". Using the quantum superposition[128] of quantum mechanics (that holds that a physical system such as a real elementary particle in this case) exists partly in all its particular, theoretically possible states simultaneously; but, when measured or observed, it gives a result corresponding to only one of the possible configurations[129] (as described

---

[127] The positron has an electric charge $(+1e)$ of 1.6×10^{-19} Coulomb, a spin of $\frac{1}{2}$, and has the same mass as an electron. When a low-energy positron collides with a low-energy electron, annihilation occurs, resulting in the production of energy

[128] P.A.M. Dirac (1947). *The Principles of Quantum Mechanics (2nd edition)*. Clarendon Press

[129] The general principle of superposition of Dirac for quantum mechanics applies to the states [that are theoretically possible without mutual interference or contradiction] ... of any one dynamical system. It requires us to assume that between these states there exist peculiar relationships such that whenever the system is definitely in one state we can consider it as being partly in each of two or more other states. The original state must be regarded as the result of a kind of superposition of the two or more new states, in a way that cannot be conceived on classical ideas. Any state may be considered as the result of a superposition of two or more other states, and indeed in an infinite number of ways. Conversely any two or more states may be superposed to give a new state. The non-classical nature of the superposition process is brought out clearly if we consider the superposition of two states, A and B, such that there exists an observation which, when made on the system in state A, is certain to lead to one particular result, a say, and when made on the system in state B is certain to lead to some different result, b say. What will

122

in interpretation of quantum mechanics). Returning to the Casimir effect the virtual particle at the beyond state exists and corresponds to a proven idea and as physical system in the *beyond* it exists as one of the possible configurations of it. This is the configuration before it bounces to real when it acquires another of its possible configurations.

The infinite virtual particles that constitute the F_c/A are proven, existing (εἶναι) and real corresponding to virtual particles that are provably defined. These concrete virtual particles correspond to ideas within the logos that describe each virtual particle exactly. Each one of the virtual particles when bounced to real will create time-space as well as all the other properties that describe it following the laws of Science thus creating one Universe for itself. Each on these Universes is completely independent from all the other ones and there is no overall time-space that links them between them. The *beyond*, as repeatedly discussed, has a space and time value of zero.

In this step observance is made only to the one virtual particle that bounced to real and triggers the creation of the Big Bang and the evolution. The *beyond* in this context is spaceless by definition however every single virtual is compatible with the definition given on the *beyond* state and becomes observable as soon as it is bounced to real.

Fifth step: from real particle to a Universe

By definition in this book a "α" particle is one that in an evolutionary process reaches at least to the point that a human mind appears. The successful absolute endpoint (tecmor) is defined when an evolutionary process reaches the point to evolve to a superlative mind.

Partial endpoints are the points when an advanced or a collective mind is achieved. The proven, real and observed endpoint is the collective mind of the Humans on earth in 2013.

In Plate 26 we define such a potential particle as "α" iconic real particle that derived from the "α" iconic virtual particle. At the instance of the

---

be the result of the observation when made on the system in the superposed state? The answer is that the result will be sometimes a and sometimes b, according to a probability law depending on the relative weights of A and B in the superposition process. It will never be different from both a and b [i.e, either a or b]. The intermediate character of the state formed by superposition thus expresses itself through the probability of a particular result for an observation being intermediate between the corresponding probabilities for the original states, not through the result itself being intermediate between the corresponding results for the original states.

start of the time (when T is greater than zero) then we define the particle as "α" primary real that will evolve in its own newly created Universe towards the "α" Primary scalar field and so forth to the Big Bang.

The time in conventional time between the "α" primary real and the "α" Primary scalar field is by definition in this book Planck's time (5.39×10^{-44} seconds). Eventually it will then evolve further to a "α" Universe explosion that will lead to all the known process even to up to a superlative mind as discussed earlier.

In what concerns the bounced from *beyond* "α" real particle from now onward, it is indifferent and detached from the *beyond* and from any other potential virtual particle that has bounced and exists now on (εῖναι) in its own proper Universe.

As with every bounced from virtual to real particle also at the instance when the "α" virtual particle bounces to the real one, it initiates the start of the timing in conventional time that will last a defined period Δt during which the particle will remain real.

The initiation of the Big Bang with the establishment of the Universe of today, according to Andrei Linde[130], starts with a scalar field. The scalar field[131] contributes in the understanding of the logarithmic expansive inflation from the start of the Big Bang in the first instances to 10^{-44} seconds.

---

[130] What Energy Drives the Universe? Andrei Linde, Department of Physics, Stanford University, Stanford, CA 94305, USA

[131] http://en.wikipedia.org/wiki/Scalar_field

Mathematically, a scalar field on a region U is a real or complex-valued function or distribution on U. The region U may be a set in some Euclidean space, Minkowski space, or more a subset of a manifold and it is typical in mathematics to impose further conditions on the field, such that it is continuous or often continuously differentiable to some order. A scalar field is a tensor field of order zero, and the term "scalar field" may be used to distinguish a function of this kind with a more general tensor field, density, or differential form. Physically, a scalar field is distinguished by having units of measurement associated with it. In this context, a scalar field should also be independent of the coordinate system used to describe the physical system—that is, any two observers using the same units must agree on the numerical value of a scalar field at any given point of physical space. In physics, scalar fields often describe the potential energy associated with a particular force. The force is a vector field, which can be obtained as the gradient of the potential energy scalar field

Sixth step: from real particle to scalar field

The chaotic inflation (expansive explosion), according to Linde[132], starts in a minimal area of 10^{-33} cm, that contains a considerable and homogeneous scalar field. This field disposes a mass (energy) smaller than one milligram. Immediately after, this area in the conventional time from 10^{-43} to 10^{-35} seconds explodes and acquires a total energy (equal at least to that of the totality of the particles that exist in the visible Universe). Today we can observe only some of these particles because most of them are already far away in the frontiers of the expanding Universe. This peculiar phenomenon occurs because the active pressure of the scalar field is negative, so the total energy of matter increases during the inflation ($dE = -pdV > 0$) gigantically.

The primary scalar field can start in a tiny domain[133] of the smallest possible size (Planck length 10^{-33} centimeter) at the largest possible density (Planck density 4×10^{94} grams per cubic centimeter). The total energy of radiation in the Universe today is greater than 10^{53} grams. According to Linde, the initial emergence of the 10^{53} gram is a simple consequence of the quantum mechanical uncertainty principle[134]. Linde states *"with a tiny domain of the smallest possible size and at the largest possible density ... the total energy of matter inside such a domain is approximately 10^{-5} grams ... then inflation makes this domain much larger than the part of the Universe we see now ..."*.

Seventh step: evolutionary process

As discussed earlier, the existence of the bounced to real, and by now material particle, initiates the first snapshot occurring at the zero starting time of the Universe. This moment also defines the start of the Δt time when the virtual particle is real that is material.

On the contrary, for the *beyond* state, this moment is static and unmovable. During this Δt time when the particle is a real material, the laws of Science will be valid and the Δt time lapse will have a length of a Planck's time or multiples of it. The duration of the Δt time cannot be less than the Planck time but it can last multiples of Planck's time up to infinity.

---

[132] In comparison observation of a single electron in a Penning trap shows the upper limit of the particle's radius is $10-24$ meters.

[133] http://energy.nobelprize.org/presentations/linde.pdf

[134] http://www.mpa-garching.mpg.de/lectures/Biermann_07/LindeLecturesMunich1.pdf

In summary the *beyond,* created from an iconic virtual to a real particle follows the quantum mechanics uncertainty mechanisms[135]. At the instance of the bouncing of the virtual particle, time is zero; hence, ΔE is infinite and completely undefined. From that instance onwards, the newly bounced primary (real) particle creates time, space and all the related variables that are valid in the Universe under the rule of the laws of Science. In the first Planck time, the ΔE is logarithmically reduced and consequently as $t>0$ to the equivalent mass will be also reduced. As soon as the critical mass of the primary (real) particle[136] is attained, then according to Linde the gigantic inflation (explosion) of the Big Bang as described by the Standard model will start. This peculiar phenomenon occurs because the active pressure of the scalar field is negative, so the total energy of matter increases during the expansion.

The immaterial consists of the iconic virtual particles in the *beyond* as defined earlier up to the point of the bouncing to a material particle. Potentially the bouncing occurred from one to an infinite number of times and initiated an unlimited number of one or different material particle types (α β, γ, δ ... μ to ν) bouncing from their respective virtual ones. We name in this context such concurrently created Universes as coverses (Plate 26).

The process of the bouncing is autogenerated and automatic without external intervention. In the Universe conventional time, the coverses translate to a single-shot that coincides with a primary scalar field just before their own independent Big Bang and Universe expansive inflation occurs. Therefore, these concurrent α, β, γ, δ ... μ to ν virtual particles potentially can lead to one or innumerable coverses that each independently will evolve possibly to a defined individual Universe. Randomness is the governance in each case of coverse. The same randomness governance was valid in the Universe where the Earth is at its initial automatic creation.

PROPOSED AUTOMATIC CREATION PROCESS

Resuming the previous paragraphs, we propose that among the possible virtual particles at some point one particle bounces to material disposing the specifications of the "primary" particle that will create the primary

---

[135] Heisenberg's equations: $\Delta p \, \Delta q > h \, / \, 4\pi$ and $\Delta E \, \Delta t > h \, / \, 4\pi$

[136] Containing a considerable and homogeneous scalar field in a minimal area of $10-33$ cm disposing a mass (energy) of the order of one milligram

scalar field and thereon the vector field and finally a post-Big Bang Universe. Plates 25 and 26 show the process leading to the creation of such a Universe. This process starts from *beyond* and the iconic virtual particle, to the appearance of virtual particles that will potentially bounce to a real.

As an example, we take a concrete virtual particle α and we name it primary virtual particle. This particle is concurrent with other potential virtual particles (β, γ, δ ... μ, ν) all of them potentially able to evolve from primary virtual to respective real particles and converses. All α, β, γ, δ ... μ to ν and so forth are scientifically provable ideas.

The bouncing creates a state compliant to the rule of the laws of Science existing in the Universe. The moment of the bounce of the iconic virtual particle to a material particle, is an achieved action and is simultaneously the start of the existence of a Universe with defined properties and compliant to the laws of Science. The bounce from virtual to real since it is a proven scientifically action, it constitutes automatically a snapshot that has occurred. Such a snapshot can have occurred at least once (the proof is by observance the existence of the actual Universe) evidence of which is the existence of the Universe where the Earth is. This bouncing from the *beyond* and the creation of a new Universe and snapshot, potentially can occur an infinite number of times.

Questions on the timing and the bouncing of the primary real particle

The *beyond*, as repeatedly mentioned, is free of time; therefore, the question that arises is when the primary real particle that evolved to the primary scalar field occurred. The answer is, in *beyond* terms, that it was a one and for all event that provably occurred at least once and this was for the Universe where Earth is. The proof of this is the existence of evolution that we observe from the Big Bang up to the human mind on Earth. In Universe conventional time, this event occurred practically at the instance of the Big Bang 13.8×10^9 years ago. Staying in conventional Universe time, the same event (the bouncing to a real particle that has the capacity to create a Universe from the respective scalar field) can have occurred more than once with the creation of up to infinite other completely independent Universes.

A relative question is how the timeless dimension of the *beyond* corresponds to the "a" particle newly created Universe time dimension. The answer is that the *beyond* is time wise static and unmovable on the contrary the Universe has a time start point, which coincides with the Big Bang and time continuity in Planck's time multiples, which is infinite∞.

127

In such a case the Universe observance of the *beyond* time is perpetual and endless despite that the *beyond per se* has no time dimension.

Another question is if the other than the "α" particle (β, γ, δ … μ to ν etc) will create Universes similar to the one where the Earth is. The answer is positive; from the instance of the creation of the real bounced particle, the evolution will follow the same laws of Science, however creating respective Universes that are completely independent than the one where the Earth is. Each of these potential Universes is rules by the laws of Science and following the governance of determinism, indeterminism and randomness as described in the earlier chapters.

A further question is if the potential Universes deriving from the "α, β, γ, δ … μ to ν" etc particle will be identical; the answer is negative since the governances (deterministic, indeterministic or randomness) will most probably evolve differently for each developing Universe (Plate 23).

Ontology of the Universe automatic creation

The ontological confirmation of the existence of the Universe, including the Earth, is the proof that creation in fact occurred at least once: the factual confirmation of this is by the observation and study of the natural history. Theoretically, this creation is not a unique event; it has occurred at numberless concurrent events each being similar or different.

Here we propose the automatic creation process from the *beyond* and from a primary particle. Based on the observation of the evolutionary pathway (Plate 1 and 2 and Plate 22 to 23) up to the human mind, the options for a successful pathway different from the one pursued is improbable but still conceivable. This is due to the evolutionary restrictions imposed by the governance and by the interplaying five variables that lead to the cosmological evolution to the planets and finally to the mind.

Beyond, governance and "primary scalar field"

The governance that commands the bouncing is randomness: this is because the uncertainty principle is valid. The virtual particle (iconic virtual particle) bouncing from the no space-time to the real particle is a static and unmovable phenomenon leading to an active action momentum creating space, time and movement. Under such circumstances, the bounce is an achieved action that is neither under determinism nor indeterminism governance. It is not under determinism since there is no prior snapshot: the *beyond* has no time dimension. It is also not under indeterminism since there is no human mind as the causative mover, the *beyond* being unmovable.

In both the zero-point energy and the *beyond* states the idea and the iconic virtual particle are immaterial compliant to the laws of Science[137] including the quanta mechanics and the Heisenberg uncertainty principle. The uncertainty principle is valid in the case of the *beyond* only after the start of time as a real particle that is after the bouncing when randomness and determinism are initiated. From the point that the real particle is bounced, the governance that started as random (Heisenberg's uncertainty principle) from now on is deterministic at the Δt interval with the initiation of a snapshot, compliant to the rule of the laws of Science.

Restrictive evolution

Plate 27 shows the evolutionary process from the *beyond* down to the human mind. The interplay of the five variables starts with a limited multiplicity, complexity, diversity, plasticity and adaptability in the first instances of the primary particle, the primary scalar field and the inflation after the Big Bang. The Universe as seen on the Earth 1.3×10^{10} years later ends to an innumerable increase of all the snapshots that evolved under the governance command and the interplay of the five variables canalizing evolution by restricting the possible pathways (as shown in Plate 23). The restrictions starting from the primary elementary particle, the atom, molecule, the biological genetics and the environmental conditions influence the endpoint, which is proposed in this book as the superlative mind.

The arrow in Plate 27 indicates the evolution from the primary particle down to the human mind that follows this restrictive pathway where the interplay between the five variables, maximize as we approach the emergence of Humans.

The noocentric evolution cannot differ significantly from what we observe on Earth today because the probability of reaching a viable Human significantly different from Humans of nowadays is improbable due to the evolutionary selective limiting restrictions. Even under different conditions in an exoplanet, the evolution toward an intelligible being much different from Humans is for the same reasons unlikely; without excluding that insignificant anatomical and physiological

---

[137] The laws of Science are valid in the *beyond* as immaterial notions (ideas), however their applicability is nil since there is no object for their appliance

differentiations can occur. The basis of this is the knowledge that we dispose at this point in the observable Universe and our profound knowledge of the natural history and biology of the evolution on Earth.

Superlative mind creation of Universes

The bouncing can also occur by an interventional, intentional and voluntary programmed external action as an action of a mind. Theoretically, the superlative mind can setup, organize and perform a series of actions that can lead to the generation of material and real particles experimentally. Keeping in mind the levels of the mind, the superlative mind can initiate a serial chain of reactions that can mark the creation and generation of a *de novo* Universe that we named in an earlier chapter as "techno-verse".

In the specific case of the fabrication of a new mind-created Universe, the process followed is the same as the one described. This state opens the potential of an interventional creation of a Universe by a superlative mind or other human mind. More precisely in laboratory conditions, a primary virtual particle creates a techno-verse *de novo* in this case under indeterministic governance. The indeterministic intervention can occur at the level of the iconic virtual particle bouncing or at any other point in the evolution.

Plate 26 again, shows the diagram of the proposed evolution from the *beyond* to the mind, an extension of the superlative mind is marked as the techno-verses. By saying this, theoretically it cannot be excluded that even the Universe where Earth lies is possibly superlative mind-made. In the same Plate is shown as a ramification of the "α" Universe by the creation of bubble multi-verses.

Top quark

We have discussed earlier the potential creation of a Universe from a *beyond* bounced electron or a positron. Another hypothetical example of an elementary particle that could represent the primary particle is the top quark that is a fundamental material constituent in the Universe. Like all quarks, the top quark is an elementary fermion with spin-1/2, and experiences all four fundamental interactions: gravitation, electromagnetism, weak interactions, and strong interactions. It has an electric charge of +2/3 e, and is the most massive of all observed elementary particles having a bare mass of 172.0 ± 2.2 GeV/c2. The antiparticle of the top quark is the top anti-quark, the anti-top that differs from it only in that some of its charge that has equal magnitude but of opposite sign.

Outside the top quark, the other particles (Plate 3, down) or other still undisclosed ones potential primary elementary particle, could bounce to real and trigger the creation of a Universe through a primary scalar field. The discovery of heavier particles disposing of a mass that approaches the minimal mass required by the Linde theory mentioned earlier will bridge the automatic creation requirements.

Whichever the virtual particle is that becomes real; its respective snapshot initiates time and space. The immediate next snapshot will follow at a chronological interval that will initiate the chronological registry of the first monoclonal snapshot series. The proposal in this book is that the Universe develops starting from the creation snapshot of the virtual to real particle: the next snapshot follows this at an interval exactly equal to a Planck's time.

From this point on the evolution will follow in at least the four known dimensions that are incompressible and indivisible. These same dimensions as measurable multiples of the Planck's units of time and length will accumulate infinitely, indefinitely and illimitably in conventional Universe terms.

The first snapshot marks the physical existence of matter and the equivalent energy. After a Planck's time follows, the first evolutionary step theoretically is the Big Bang. From thereafter the evolution follows the well-described cosmological model in a chronological order of sequential snapshots.

At the end of the first snapshot, the Universe consists of a real particle that possesses mass, volume and gravity. The second snapshot that follows possibly is the start of the inflation, created by the instability of the first particle.

THE UNMOVED MOVER OF ARISTOTLE

The *beyond* as described here is compatible with the Aristotelian unmoved mover and linked to what Aristotle coined as the first philosophy[138].

---

[138].. ἔστι τοίνυν τι καὶ ὃ κινεῖ. ἐπεὶ δὲ τὸ κινούμενον καὶ κινοῦν [καὶ] μέσον, τοίνυν ..

"Then if there is not some other substance besides those which are naturally composed, physics will be the primary science; but if there is a substance which is immutable, the science which studies this will be prior to physics, and will be primary philosophy, and universal in this sense, that it is primary. And it will be the province of this science to study

"... and since that which is moved while it moves is intermediate, there is something which moves without being moved; something eternal which is both substance and actuality ..."

Plate 28 shows the Universe from creation to the emergence of the human mind separated by two distinct states: the immaterial *beyond* and the material Universe that followed. Both the *beyond* and the unmovable mover are compatible with immaterial notions. In the case of the unmovable mover, it is *"... immutable and universal in this sense"*.

In this Plate the verified ideas, are compatible with and encompass the definitions of both the *beyond* and the unmovable mover. The *logos* is compatible to both the *beyond* and the unmovable mover since it is immaterial, immutable and is compatible with the immaterial notions. In the same immaterial state is included *dynamis* that represents the potential of several action or movement options not yet initiated or set into the process of actualization as well as the virtual particle of the *beyond*.

However the *beyond* and the *logos* are incompatible with the universals of Aristotle that are types, properties, or relations that are common to their various instances and exist only in things (material entities) and never apart from them. In this book the Platonic definition of ideas is followed.

The Plate splits in two columns, on the left it refers to the Aristotelian terminology and on the right to the creation proposed in this book.

On the Aristotelian terminology, the definition of immovable mover refers to the passage stating that there is something, which moves without itself "being moved". We extrapolate to the *beyond* as described earlier saying that it is analogous to the immovable mover. In turn, the real elementary particles after the bouncing from virtual are effectively of substance (material), belong to the Universe and correspond to the Aristotelian "movable" (κινούμενον).

- Therefore, the immovable mover is the primary cause of movement of either a potential (εν δυνάμει) and or of an achieved action (εν ενεργεία).

- The immovable mover is motionless and hence it is *per se* with potentiality (δύναμη) but without actuality (energy, ενέργεια).

- The first cause of an entity is a force (δύναμις) as a potential of being at work or activated to move.

Being qua Being; what it is, and what the attributes are which belong to it qua Being".
Perseus. Aristot. Met. 6.1026a

- Therefore, the immovable mover being immobile has potentiality (δύναμη) for action (ενέργεια).

- Entelechy (εντελέχεια) is the completeness of action (ενέργεια) that is the achievement of the action of any entity in its being-at-function.

- The immovable mover has the potential of actualization of an action or movement. The achievement of the completeness (εντελέχεια) is when the material "thing" (τι) is finally moved (κινηθεί).

This Plate 26 shows the comparability of the two lines derived from either the unmovable mover or the *beyond* that lead to an evolution ascending to a superlative mind.

Limitations of theory

The proof that the evolution from the first instances of the Big Bang to the human mind is an objective provable event is self-evident based on the study cosmology, natural history on Earth and anthropology. The theory described in this book for the automatic creation and evolution to the human mind has two limitations: the first refers to the bouncing to a real particle in the state of the *beyond* and the second the primary scalar field required for the Linde inflation. Theoretically the three principles base the automatic creation hypothesis launched here.

Related to the *beyond* the bouncing to the real particle is based axiomatically to the extrapolation of Dirac's general principle of superposition of quantum mechanics, the Dirac delta function and to the uncertainty principle of Heisenberg. Related to the inflation after the Big Bang the mass of the primary virtual particle that initiates the creation is not quantified in order to reach the approximate 10^{-5} grams.

However both these limitations require specific experimental proof.

BIBLIOGRAPHY

Aristotle Bilingual Anthology. Elpenor. The Greek World.
http://www.ellopos.net/elpenor/greek-texts/ancient-greece/aristotle.asp
Ancient Greek Music at the Commission for Ancient Literature of the
Austrian Academy of Sciences. 2004.
Anthology of Plato & Aristotle.
Beazley Archive. Publications.
http://www.beazley.ox.ac.uk/Publications/Script/SHCVolume1Contents
.htm
Burcham and Jobes. 1995. Nuclear and Particle Physics. Longman
Scientific & Technological.
Chalmers, David John. 2003. Philosophy of mind: classical and
contemporary readings. Oxford University Press.
Chandra X-ray Observatory. CXC operated by NASA.2006.
http://chandra.harvard.edu/
Copleston, Frederick. 1993. A History of Philosophy, Vol. 1: Greece and
Rome. From the Pre-Socratics to Plotinus. Image Books.
Darwin, Charles. 1861. On the origin of species by means of natural
selection,: Or, The preservation of favoured races in the struggle for life.
John Murray.
Dictionary of the Philosophy of the Mind. Eliasmith, D. Editor. 2006.
Goldsmith, D. and Owen, T. 1993. The search of life in the Universe.
Addison-Wesley Publishing Company.
Guth, Alan. 1997. The Inflationary Universe. The quest for a new theory
of cosmic origin. Helix Books. Perseus Books.
Hocking, WE. 1929. Types of Philosophy. Charles Scribner's Sons.
Kirk, GS, Raven, JE, Schofield, M. 1990. Presocratic Philosophers.
Cultural Foundation of the National Bank of Greece.
Linde, A. D. 1990. Inflation and quantum cosmology. Academic Press,
Boston and A. D. Linde. 1990. Particle physics and inflationary
cosmology. Harwood Academic Publishers.
Linde, A. D. 2005. Towards Inflation in String Theory. Journal of
Physics: Conference Series 24: 151–160.
Linde, A.D. 1990. Particle Physics and Inflationary Cosmology
(Contemporary Concepts in Physics Series). Translated from Russian.
CRC Press.
Plato : Complete works. Elpenor. The Greek World.
http://www.ellopos.net/elpenor/greek-texts/ancient-
greece/Plato/default.asp

Popper, Karl, 2002. The Logic of Scientific Discovery. Routledge Classics.

Russell, Bertrand. 1961. History of Western Philosophy. George Allen & Unwin.

Russell, Bertrand. 1989. Problèmes de philosophie. Bibliothèque Philosophique Payot.

Serway, R.A. 1993. Contemporary Physics. 4th Volume . Physics for Scientists and Engineers, 3rd Edition. Saunder Golden

Silk, J. 2006. The Infinite Cosmos: Questions from the Frontiers of Cosmology. The Oxford Press.

Sparrow, Gilles and Dava Sobel. 2011. Cosmos: A Field Guide. Quercus.

Taine, H. 1889. Lectures on Art. Henry Holt and Company.

Teilhard de Chardin, P. 1956. L'apparition de l'homme. Editions du Seuil.

Teilhard de Chardin, P. 1955. Le Phénomène humain. Editions du Seuil.

Tzavara, Gianni. 1988. The poetry of Empedcles. Texts, Translation, Comments. Dodoni Publications.

Hatzopoulos, Od. Publisher. 1992. Presocratics. Ancient Greek Secretariat "The Greeks" Volumes 1 to 10.

Weinberg, Steven. 2008. Cosmology. Oxford University Press.

Wikipedia has been widely been used for definitions.

http://en.wikipedia.org/wiki/Main_Page

QUESTIONS AND ANSWERS

If zero time of the Universe is the Big Bang, what was before?

There was nothing material before

In this book the before the Universe start is described as the *"beyond* or *επέκεινα or transcendent"*

No matter, no forces, no pressure, no field, no energy, no spatial dimensions, no electrical charge, no time, no gravity, no temperature

The *beyond* is independent of any property, condition, state, dependency except its ability to create from virtual particles real particles

The ideas are compatible with the *beyond* since not requiring any property or variable in order to "εἶναι"

What happened at the instance of the start of the Universe?

A virtual particle that exists as an abstract notion spontaneously becomes real by borrowing energy in accordance with the general principle of superposition of quantum mechanics of Dirac and applies to the states that are theoretically possible without mutual interference or contradiction

This is provably occurring in the zero-point energy state in the Universe following the rule of the laws of Science and according to the uncertainty principle of Heisenberg, the superposition principle of Dirac and the Delta function of Dirac

What is the theoretical approach that backs the *beyond* approach and the creation of the Universe from a virtual particle?

It is based on the following three principles: Heisenberg uncertainty principle, Dirac's superposition, Dirac's delta function

Was the Universe an automatic creation?

Yes. A Universe like the one that contains the Earth does not require any indeterministic or causative intervention for its creation

It is created automatically by the emergence of an elementary particle out of the *"beyond"* acquiring material properties immediately at its creation, including dimension and time

The iconic virtual particle is an immaterial idea that described the properties of the real particle that is compatible to the *beyond*

The created real particle creates automatically its Universe as described in cosmology and as happened with the historical Universe where the Earth is

Is the starting real virtual particle that bounced from virtual the Higgs boson?

Probably not. In the Standard model the fermions as well as the energy carrying bosons are tetractides (foursomes)

The evolution of the Universe up to the appearance of the mind is defined as a series of evolving tetractides of four elements, four hadrons, four atoms, four molecules, four nucleotides. This observation is based on the concepts proposed by the philosopher Empedocles

The existence of a fifth boson outside the tetractide of photon, gluon, boson Z and W excludes the Higgs boson as part of the evolutionary process towards man and his mind

Which known elementary particle can be suspected of being the bounced to real creation particle

It is proposed that this might be either an electron or a positron or hypothetically the top quark

From the *beyond* virtual particle creation how do we reach to the Big Bang?

From an iconic virtual particle to the real *beyond* created particle by extrapolation similar to the Casimir effect

The iconic virtual particles are ideas scientifically provable that evolves to virtual particles following the Casimir equation

The bouncing to the real particle will create an individual Universe in each case that will follow the general the uncertainty principle of Heisenberg as well as the principle of superposition of quantum mechanics described by Dirac and the delta function of the same

The instance of the creation real particle from the iconic and the virtual is the start of relative time and will obtain minimal dimensions

What are the next events?

In the next Planck time, the uncertainty is increased as well as the ΔE following the uncertainty principle

The mass at the first Planck's time that was extremely high due to the minimal Δt will start being reduced

When the mass (energy) of the elementary particle that contains a considerable and homogeneous scalar field reaches to the point to dispose a mass (energy) of the order of approximately 10^{-5} grams, then according to Linde the gigantic inflation of the Big Bang as described by the Standard model will start

The dimensions of the newly created primary real particle will implode to a minimal of 10^{-33} cm

What happens after the implosion of the primary real particle?

The newly created Universe from the real particle constitutes a scalar field that imploded due to its self-gravitation creating a gravitational singularity

The particle tends to reach the Planck length and the Planck density

The negative gravitational energy exactly cancels the positive energy represented by the initial mass (energy) of the primary real particle

This peculiar phenomenon occurs because the active pressure of the scalar field is negative, so the total energy of matter increases during the inflation

If the Universe was created at time-zero, what were the events until 10^{-44} seconds later?

The spontaneous creation of a primary real particle from a virtual

Virtual particles as proven abstract entities constitute ideas compatible with the *beyond*

Each real particle thus created cannot render back the energy borrowed since the *beyond* from which it was automatically created has no other property than the *per se* creation of the particle

The real particle creates dimensions, time, mass, energy, fields, electrical charge at the zero time of its creation from its virtual abstract counterpart

Since the *beyond* is timeless in conventional time after each creation of a real particle an infinite number of real particles exist each constituting a potential independent Universe

Is the *beyond* compatible with abstract immaterial notions?

Yes. All numerals, forms, formulas, equations and all notions that are expressed abstractly do not require matter or any other property to exist

Therefore, following Parmenides definition they "are, εῖναι", they constitute "ideas", and since immaterial are compatible to the *beyond*

Are "ideas" infallible?

An immaterial notion is not *de facto* a verified true idea: it is a signal information message that contains all the essential data so when captured and restored, it reproduces all the contained initially information

Immaterial notions are abstract information: like numbers, digits, chemical formulas, and laboratory results, configurations of solids and volumes, mathematical relations, factors, constants, ranks of mathematics, geometric forms, musical tones, and colors among other

Only notions that are infallible, scientifically proven and real are ideas

Are abstract notions compatible with *beyond*?

The nothingness of the *beyond* is compatible with the existence of all real abstract notions (ideas)

All ideas as notions are either scientific, art, or those related to true and real states and those related to ethics and activities

Every idea corresponds to a defined numeral which in turn is equivalent to a concrete real, true and proven state (or scientific or art or ethic or activity) that "is, εῖναι" following Parmenides

Is the *beyond* an absolute vacuum?

No. The absolute vacuum is defined in the Universe in contrast to *beyond* that is timeless and was before the creation of the Universe. The *beyond* is free of energy, time, and dimension or of any other variable other than the potential for the automatic causeless bouncing from iconic virtual creation of real elementary particles

What are the particularities of *beyond*?

In this book, the *beyond* is a state without any fields, functions, variables and properties (other than the creation of real particles from virtual)

Therefore, no mass, particles, antiparticles, speed, energy, fields, radiation, gravity, forces or any other property, condition, situation or state is present

The real particle constitutes the totality of the newly created Universe having at the zero time nothing outside its own entity

Where is the virtual particle?

The virtual particle is an abstract notion that requires no other property to exist

The information notion of the virtual particle that will emerge to a real and create a Universe is by definition true, scientifically and experimentally proven and therefore is an "idea" that "is, εἶναι "as described by Parmenides and compatible with the *beyond*

How is the *beyond* linked to the creation of the Universe?

The *beyond* is extrapolated to the Casimir effect equation where space and distance between the plates are zero with consequence in creating an infinite force and an infinite number of virtual particles that potentially can bounce to real

Each such real created particles constitutes a potential Universe

The Universe where the Earth lies is described as an observance of this process

The Casimir effect can explain the automatic creation of a Universe?

Yes. The *beyond* has no dimensions and therefore in the equation of Casimir

$$\frac{F_c}{A} = -\frac{d \langle E \rangle}{da\ A} = -\frac{\hbar c \pi^2}{240 a^4}$$ α is zero as is also the surface A. By this the F_c takes an infinite negative value that is enclosed in the bouncing of a virtual to a real particle creating a new Universe

The beyond having no time, in relative Universe time is eternally creating Universes by processing virtual particles to real

What happens to the real particle that disposes infinite negative energy?

It implodes before exploding and initiating a Big Bang situation at each time and the process is repeated infinitely

Can we repeat the process in a Casimir effect model by minimizing the distance between the plates?

Yes theoretically and in the case that we do not use plates but a microscopic sphere then we can create negative energy; the amount of which is inversely dependent to the sphere diameter

In this case approaching a zero value diameter we could develop huge amounts of a negative implosive force

Is the evolution indeterministic?

Yes. However, only after the point when the human mind appears

The levels of mind are described in gradual steps based on the natural history on Earth

The evolution trend of the levels of the mind from an individual mind to a superlative mind are defined

The superlative mind has knowledge of all the "ideas" either being real states or scientific, aesthetics, ethics/activities notions

Is the evolution, since creation, ascending?

On Earth there is a historical trend observed of ascending evolution to the year 2013 notwithstanding the collective mind of the human societies is far away yet from reaching the superlative mind level

It is postulated that there is a probability that the civilization on Earth is a failure or a dead-end and does not lead to a higher mind level

The governance, the five variables (5πs) combined to environmental and historical circumstances in the Universe globally restrict the flexibility of the evolutionary process

It is most probable that the pathway the Universe evolution is the one that led to *Homo sapiens* on Earth or in other exoplanets

However, without this meaning that *H.sapiens* is the thread to the appearance, in the past or in the future, of a superlative mind

Are Humans immortal?

Reference is made to the ethics of genetics and it is postulated that the DNA as an abstract numeral notion is reproducible under specific condition and therefore immortal

However, the DNA of a Human or of an animal can be reproduced and launched to a new environment but the life cycle of this individual will be completely different from its identical twin with which he or she shares the same genetic code

Are Humans reproducible?

Yes. At the instant of the fusion of the male and female genetic information taking as example mammals, the newly created first cell of the new individual has a genetic code that is unique and is reproducible under similar conditions anywhere where such conditions are satisfied

Therefore, this initial code at the zero-time of conception is a numeral and represents an idea

From the "zero point" of conception and thereafter it is the external conditions that determine the differentiations that are occurring at each instance to the initial zero-time genetic code

Is the human mind an information system?

Yes. The human mind is described as a material information functional system formed from cellular biological infrastructure and from information notions

The information notions are conveyed and expressed electrically, physically or chemically that correspond to abstract immaterial numerical signals notions

Each notion corresponds to an abstract numerical

The human mind is material and processes only material information notions

The notions processed by the material human mind correspond either to proven true immaterial ideas or to wrong empirical ideas that are errors or fantasies

What is the definition of the mind?

The mind is a human corporal and material entity and is described in anatomical and physiological terms as to the processing of functions, programs, methods, tools and information notions

Several animalia possess sporadically certain advanced psychological and intellectual properties resembling to the mind (reflexes, responses and instinctive passive or active actions that are also shared by Humans

Animalia dispose of nervous systems that are through evolution stepwise upgrading to reach, globally, to the central nervous system of Humans

Several specific properties of certain animalia are more advanced than the respective ones in Humans, however, globally animalia are less developed as to the nervous systems

Only when the evolutionary ladder reaches the mind Humans acquire the capacity to process and prove real states as well as scientific, aesthetic, ethic/activity notions

What are the categories of material and immaterial notions?

All notions can be immaterial proven ideas or material thoughts, matrices, empiric ideas, stimuli or physical signals

The ideas as notions are categorized as true (real) states, scientific, aesthetic, ethics/activity

The mind can processes only material notions that are: thoughts, matrices, empiric ideas, stimuli, errors, fantasies, waves, signals, mechanical emissions etc

All the notions, either real proven immaterial ideas or material notions of the mind correspond to concrete numerals

Every material entity in the Universe corresponds to a concrete respective numeral that fully describes it and from which it can be reproduced when satisfactory environmental conditions will exist

All material entities and their corresponding numeral constitute a verified idea of a true (real) state notion

The gradual levels of the mind (individual, collective, advanced, superlative) can process and prove as true and real a number of material notions as well as unproven or errors or fantasies

Where are the ideas?

Ideas, as immaterial abstract notions, require no other property than the proof that they are real, scientifically proven, beautiful, justice, virtuous, wise or prudent

Ideas are compatible to the "*beyond*" or the Universe including the zero-point energy state

The different levels of the mind stepwise and gradually upgrade from an initial single idea level of the pithecanthropus or the early Humans to the superlative mind that processes and operates all existing ideas in the Universe

However, the verification and proof of a notion as true idea by any level of mind simply attributes the symptomatic correspondence of the material to the immaterial notion without this meaning that any mind level possesses immaterial properties

There is no miscibility between immaterial notions (ideas) and material

Is there determinism or indeterminism for actions?

In man the mind offers either active indeterministic actions to respond to active voluntary purposes (active Will) or passive actions (involuntary, passive Will)

Passive actions respond to biological acquired reflexes through the nervous system or through cellular stimuli-receptor reactions

Human mind responds deterministically to several biological as well as acquired religious, societal or moral impetuses and taboos

Are deterministic actions voluntary or involuntary?

Voluntary actions are responses to either active Will (being indeterministic) or to passive Will (being deterministic)

In all living organisms that do not dispose of a mind the final action taken for each snapshot of existence is deterministic dependent on both the environmental conditions and the genetic information, powered by passive Will

Some animals as well as Humans powered by passive Will have responses that are pyramidal performing motor responses

Indeterminism is initiated later in the evolution with the appearance of the human mind

For Humans the responses are either indeterministic where the mind intervenes deciding for an active action (powered by active Will) or deterministic as with all other living organisms (powered by passive Will)

How are the levels of the mind defined?

The mind level is individual, collective, advanced and superlative

The start of the mind is defined based on its evolutionary capacity to process the inaugural notion of "Ego sum Ego"

The superlative mind is on the other end having the capacity to possess knowledge of all the existing in the Universe ideas (that is all scientifically proven and true immaterial abstract notions and their correspondence to their material counterparts including all the true (real) states, scientific, aesthetic and ethics/activities notions)

Is this superlative mind identical to God?

The creation is presented in this book as being automatic and godless

In this book, the superlative mind definition includes the property of "good" as a combination covering beauty, fineness, excellence, goodness and unity.

It is postulated that at some evolutionary stage the superlative mind and within its property of "good" will be in a position to create a Universe (technoverse)

Are all the "minds" except of the superlative dead ends?

The evolutionary history of the Universe from creation to now as observed subjectively shows an ascending trend leading to the mind

Therefore, the evolutionary pathway leading to the appearance of a superlative mind is set as the utmost ascension (tecmor)

There is no guarantee that each evolutionary pathway will reach the superlative mind

However, all "failed" side pathways traced are necessary to the evolutionary process in order to selectively and by the interplay of the five variables will reach eventually to the superlative

PLATES

PLATE	TITLE
1	TEMPERATURE-TIME
2	TETRACTIDES AND UNIVERSE EVOLUTION
3	FORCES AND PARTICLES IN EVOLUTION
4	UNIVERSE, ENTITIES AND FORCES
5	PARTICLES, FORCES, EXPANSION, COOLING
6	INFORMATICS OF MIND AND BODY
7	INFORMATION LEVELS OF ORGANISMS
8	GENETIC INFORMATION
9	ASCENDING DARWINIAN EVOLUTION
10	GENETIC GENEALOGY
11	ARCHITECTURAL STRUCTURE OF MIND
12	HUMANS AND OTHER ANIMALS
13	MIND AND HUMAN BODY
14	MIND, IDEA AND LOGOS
15	NOTIONS AND MINDS
16	INFORMATION PATHWAYS IN THE MIND
17	EVOLUTIONARY PATHWAY
18	MATERIAL - IMMATERIAL INTERCHANGE
19	MIND AND BEYOND
20	ASCENDING EVOLUTION OF MINDS
21	NOTIONS CATEGORIES
22	EVOLUTION ASCENDING TREND
23	HUMAN MIND CENTERED EVOLUTION
24	MINDS
25	VIRTUALITY IN UNIVERSE AND *BEYOND*
26	*BEYOND* AND UNIVERSE
27	MIND-CENTERED RESTRICTIVE EVOLUTION
28	UNMOVABLE MOVER AND *BEYOND*
29	GENERAL THEORY OF CREATION AND EVOLUTION

PLATE 1
TEMPERATURE-EXPANSION AND TIME

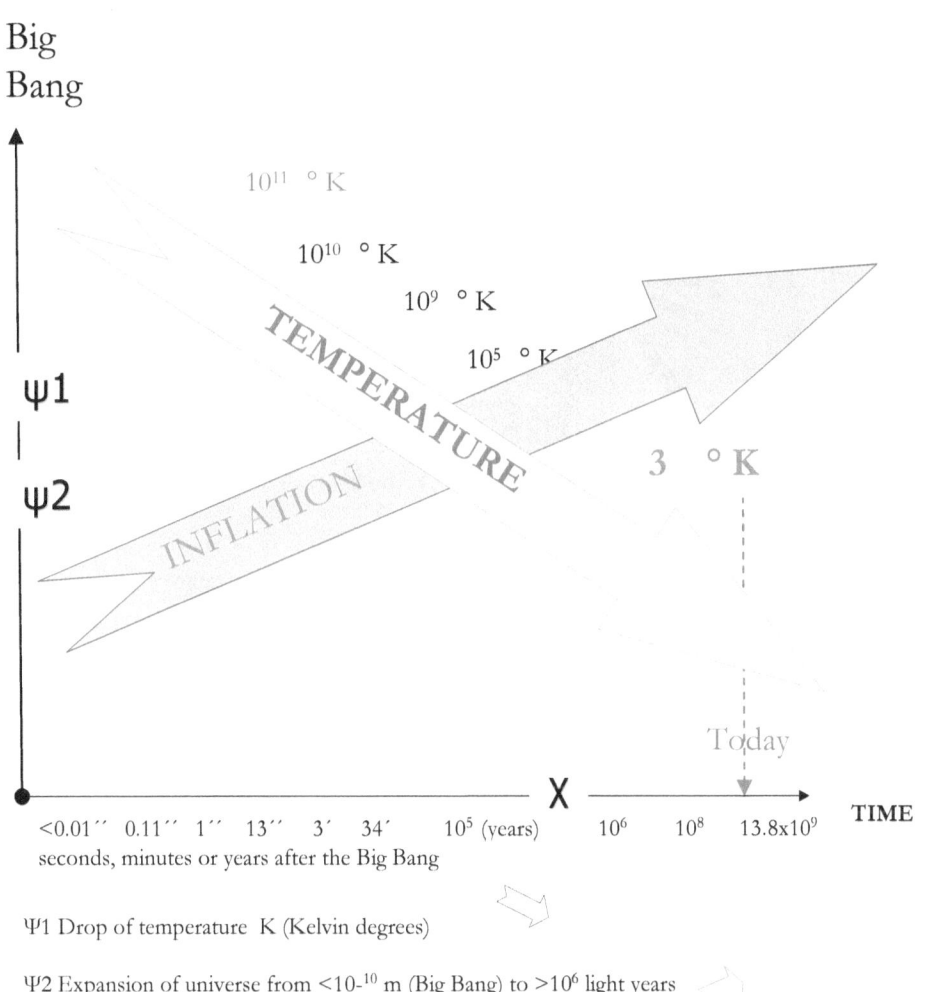

Big
Bang

10^{11} ° K

10^{10} ° K

10^9 ° K

10^5 ° K

TEMPERATURE

INFLATION

3 ° K

Ψ1

Ψ2

Today

X

<0.01˝ 0.11˝ 1˝ 13˝ 3′ 34′ 10^5 (years) 10^6 10^8 13.8x10^9 **TIME**

seconds, minutes or years after the Big Bang

Ψ1 Drop of temperature K (Kelvin degrees)

Ψ2 Expansion of universe from <10^{-10} m (Big Bang) to >10^6 light years

PLATE 2
TETRACTIDES AND UNIVERSE EVOLUTION

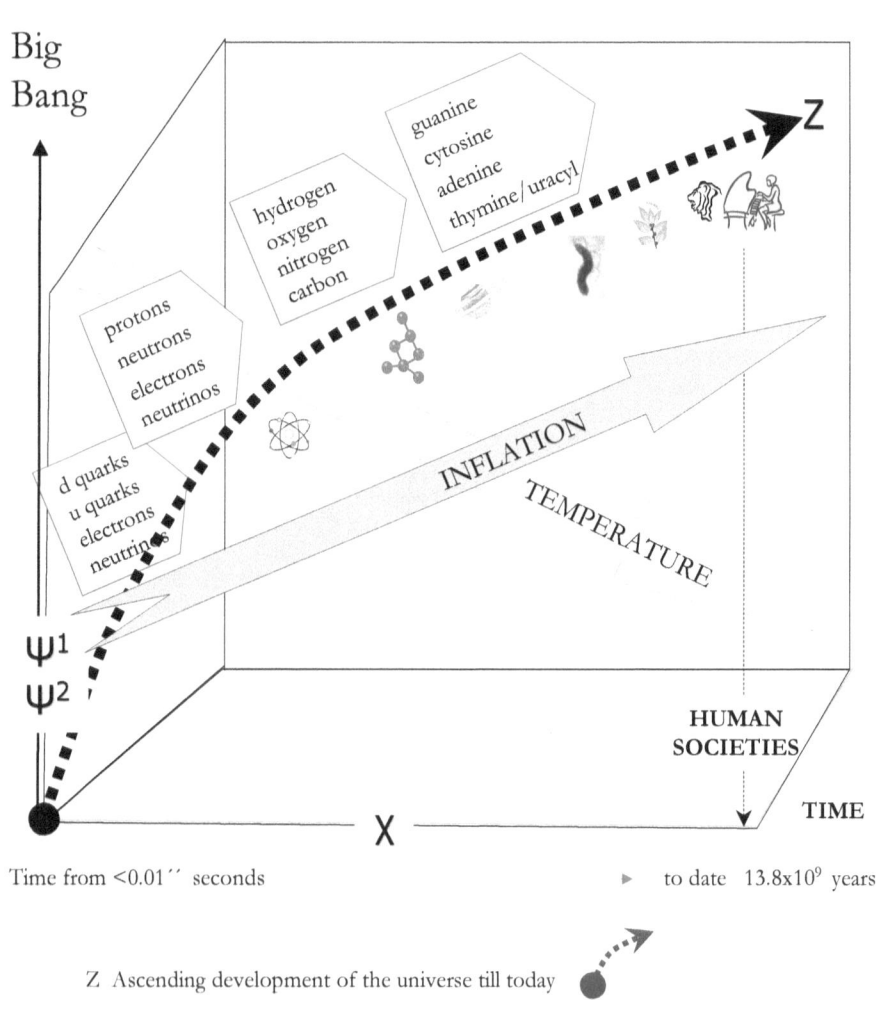

Big
Bang

guanine
cytosine
adenine
thymine/uracyl

hydrogen
oxygen
nitrogen
carbon

protons
neutrons
electrons
neutrinos

d quarks
u quarks
electrons
neutrinos

INFLATION

TEMPERATURE

Ψ1
Ψ2

HUMAN
SOCIETIES

TIME

X

Z

Time from <0.01´´ seconds ▶ to date 13.8x10⁹ years

Z Ascending development of the universe till today

Ψ1 Drop of temperature K (Kelvin degrees)

Ψ2 Expansion of universe from 10⁻¹⁰ m (Big Bang) to 10⁶ light years

PLATE 3
FORCES AND PARTICLES IN EVOLUTION

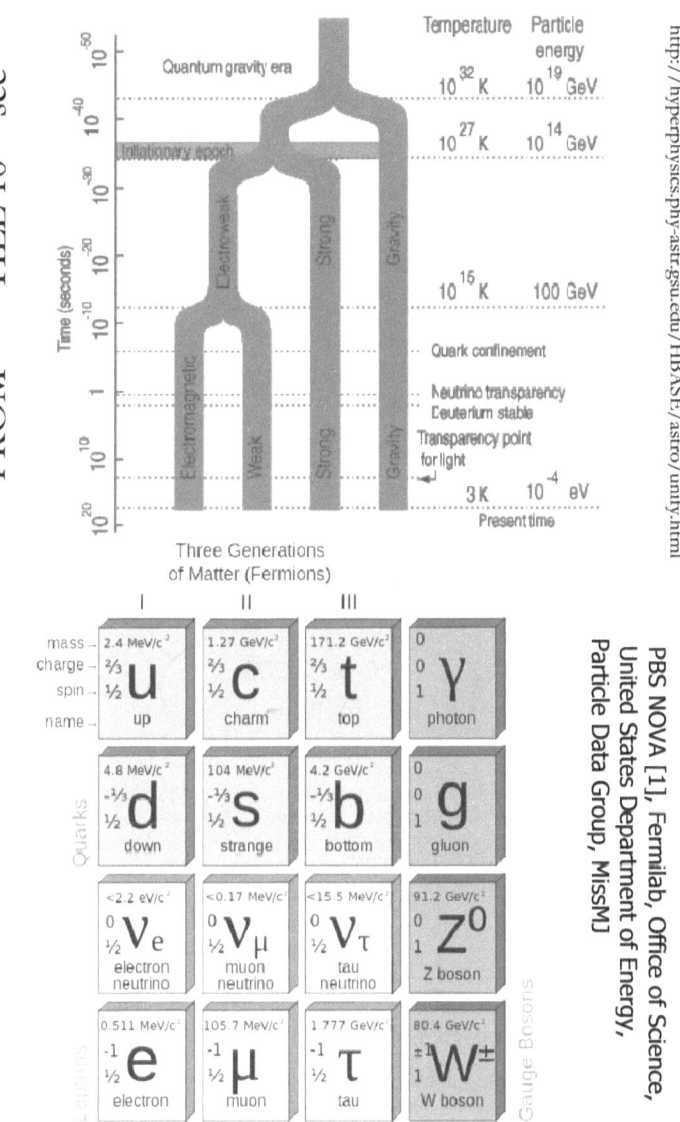

PLATE 4
UNIVERSE, ENTITIES AND FORCES

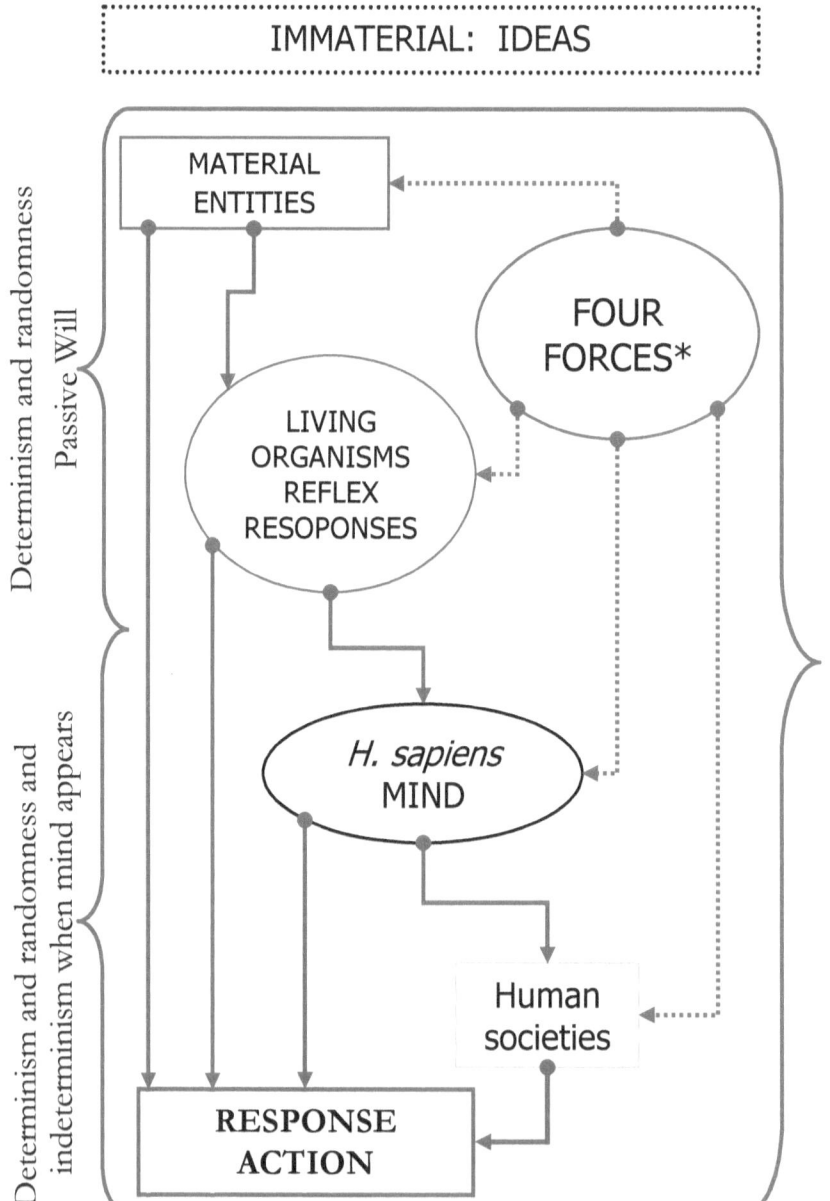

* Four fundamental forces

PLATE 5
BIG BANG SIGNAL TETRACTIDES

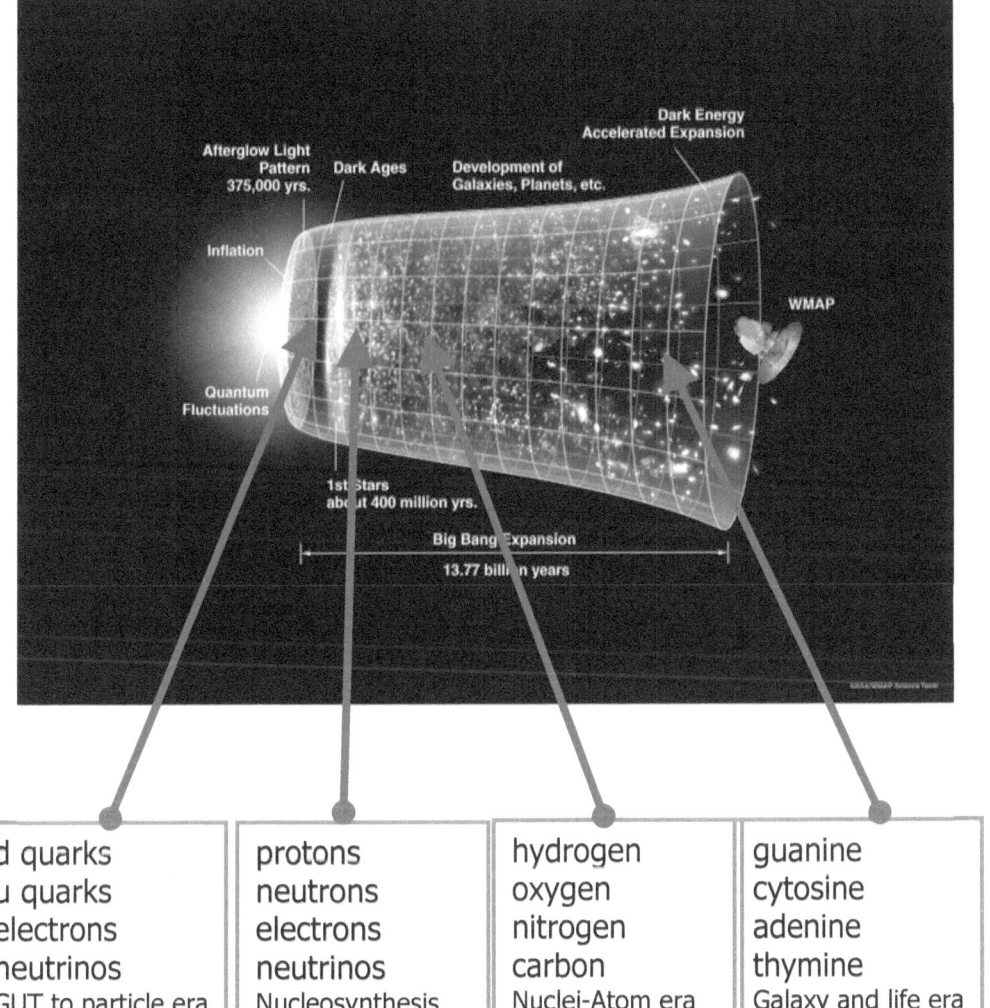

d quarks	protons	hydrogen	guanine
u quarks	neutrons	oxygen	cytosine
electrons	electrons	nitrogen	adenine
neutrinos	neutrinos	carbon	thymine
GUT to particle era	Nucleosynthesis	Nuclei-Atom era	Galaxy and life era
10^{-43} to 0.001 sec	0.001 sec to 3 min	3 min to 10^9 years	10^9 to date

http://map.gsfc.nasa.gov/media/060915/index.html

PLATE 6
INFORMATICS OF MIND AND BODY

	MIND*	BODY
ASSIGN FUNCTION	Aesthetic Intellectual Psychological	Sensory, Respiratory, Cardiovascular Urogenital, Digestive, Nervous System, etc
RUN PROGRAM	Conception Response Perception	Reproduction Homeostasis
APPLY METHODS	Inspiration Management Implementation	Movement Metabolism Communication
INFORMATION NOTIONS	Thoughts Matrices Empirical idea	Stimuli

It is postulated in this book, that both the mind and the body share the same infrastructure of cells, tissues and organs.

The mind is defined as an information system processing notions.

*The mind is graduated in levels (individual, collective, advanced, superlative)

PLATE 7
INFORMATION LEVELS OF ORGANISMS

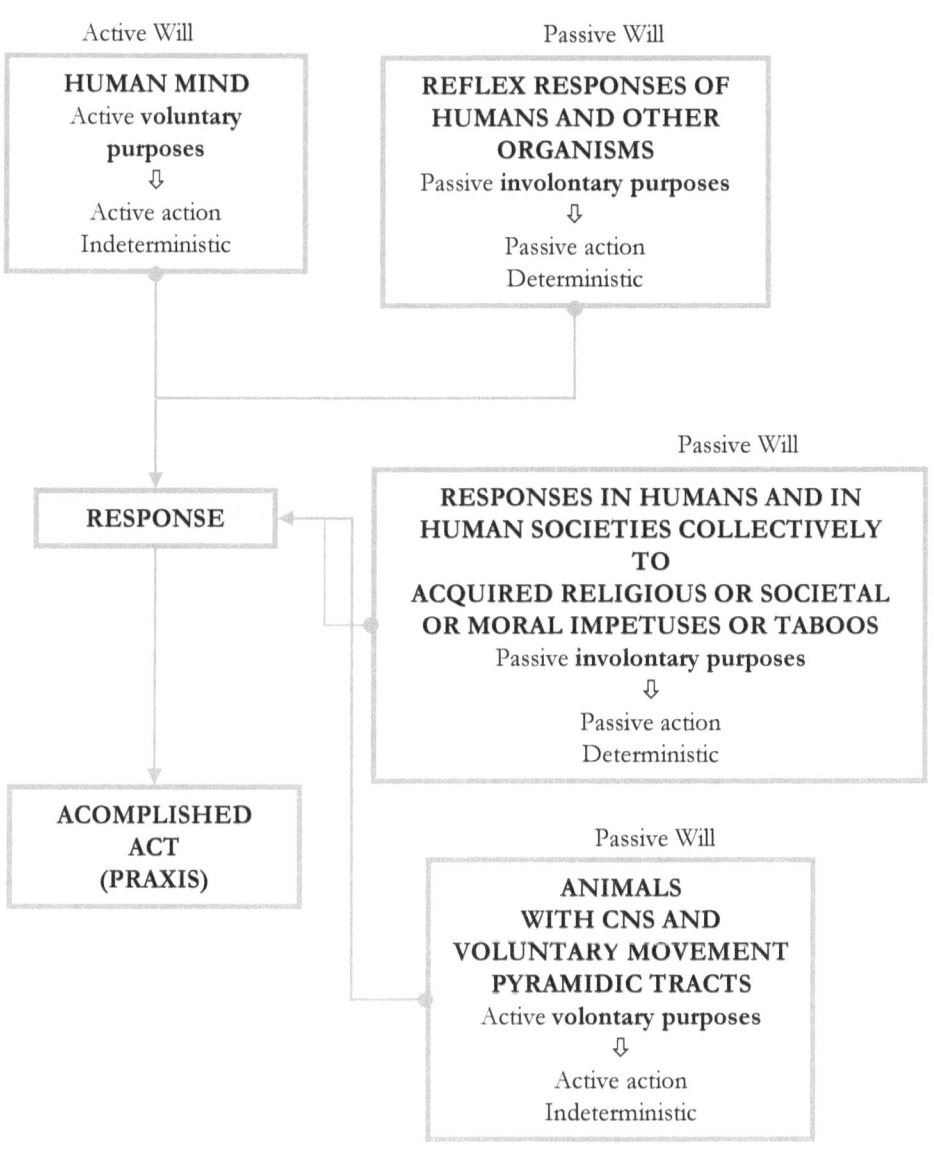

PLATE 8
GENETIC INFORMATION

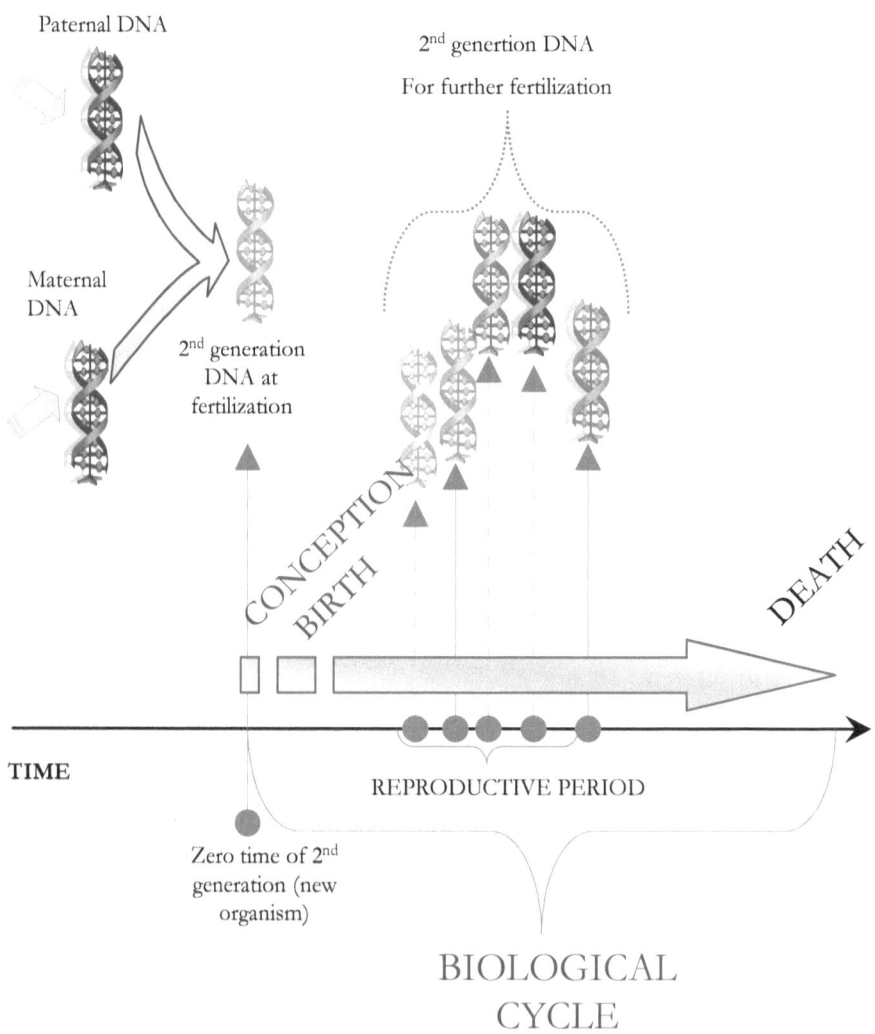

Environmental effect at all times

PLATE 9
ASCENDING DARWINIAN EVOLUTION

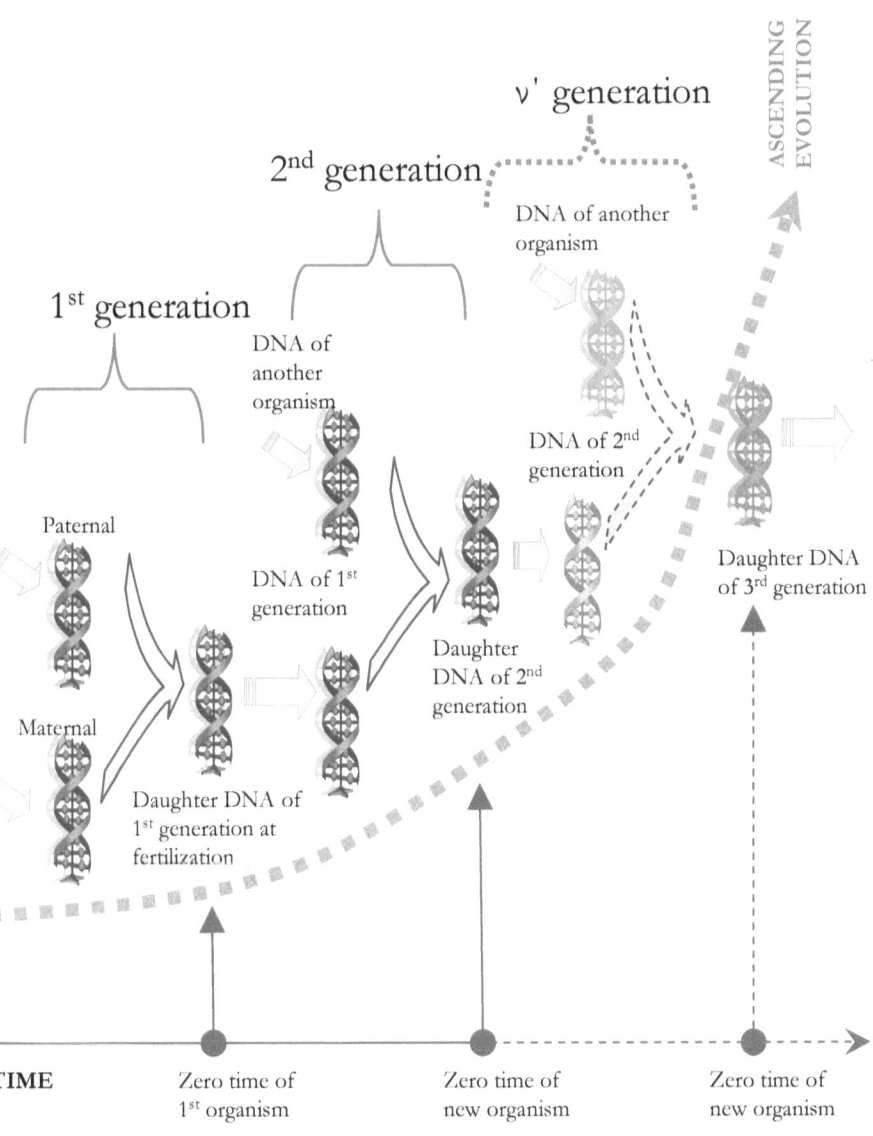

ASCENDING EVOLUTION

ν' generation

2nd generation

DNA of another organism

1st generation

DNA of another organism

DNA of 2nd generation

Paternal

DNA of 1st generation

Daughter DNA of 3rd generation

Daughter DNA of 2nd generation

Maternal

Daughter DNA of 1st generation at fertilization

TIME

Zero time of 1st organism

Zero time of new organism

Zero time of new organism

Environmental effect at all times

PLATE 10
GENETIC GENEALOGY

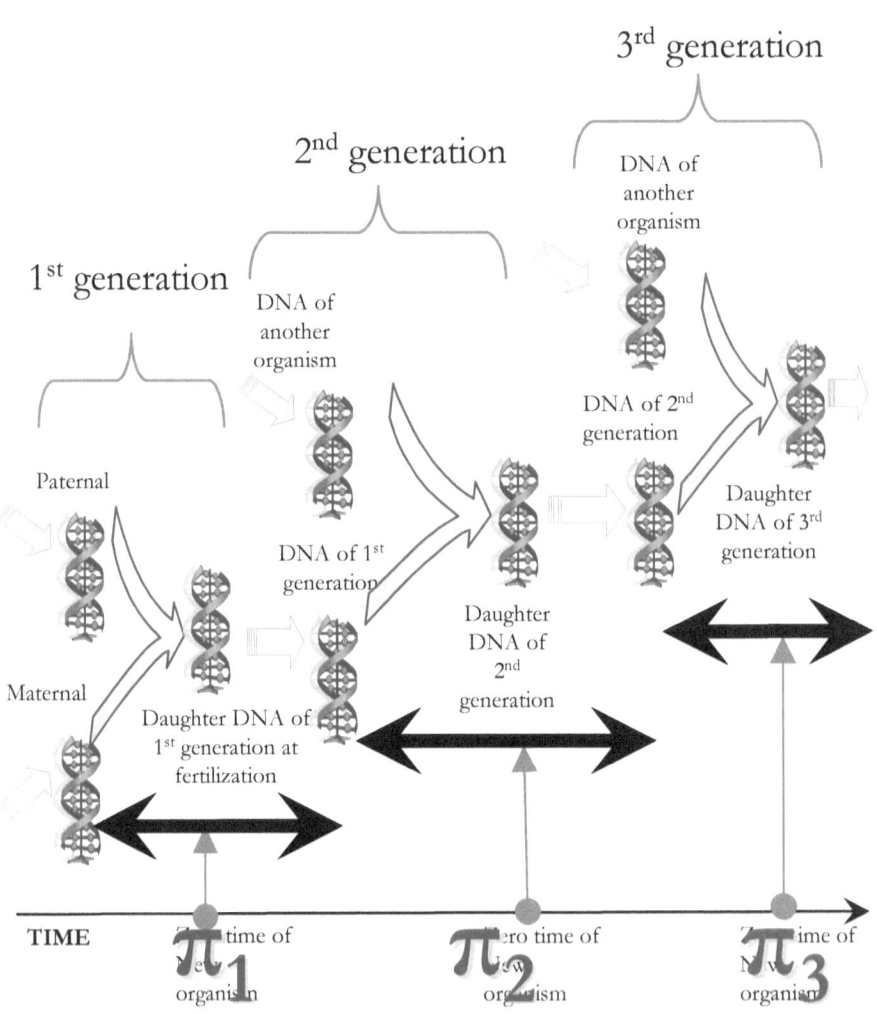

Environmental effect at all times

PLATE 11
ARCHITECTURAL STRUCTURE OF MIND

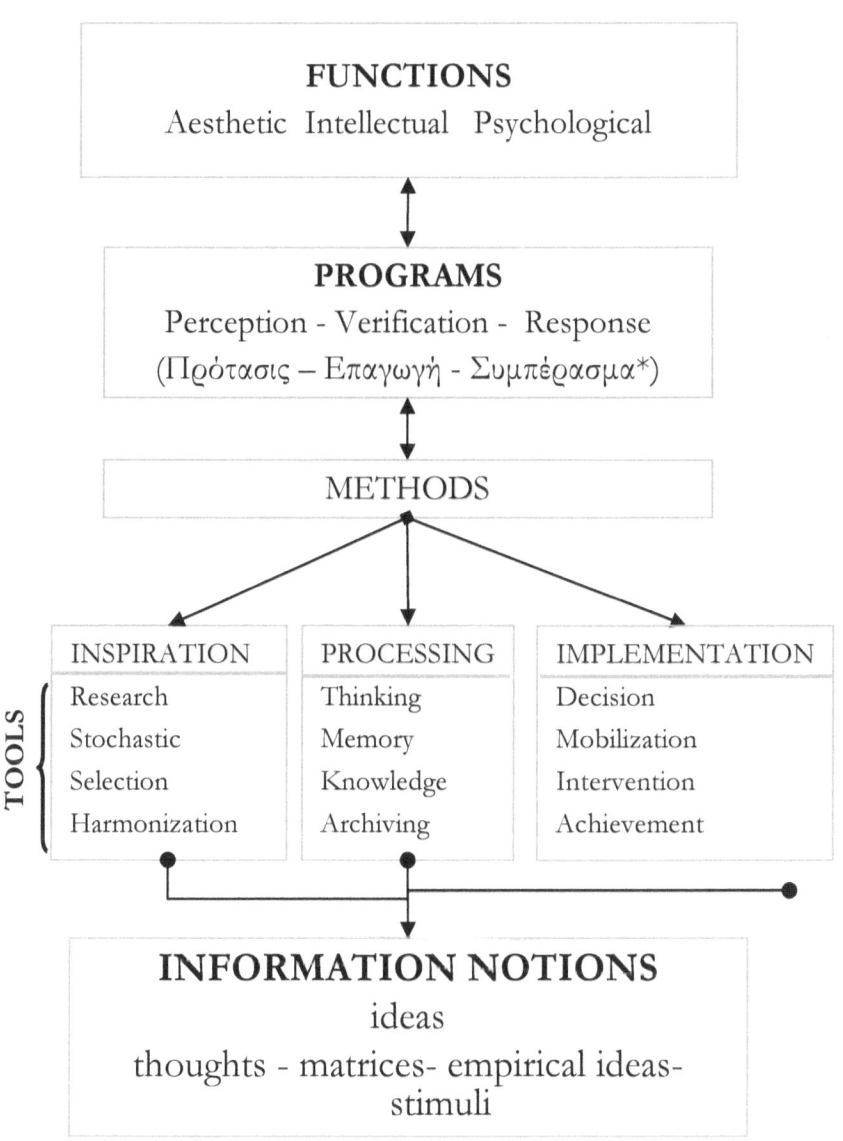

FUNCTIONS
Aesthetic Intellectual Psychological

PROGRAMS
Perception - Verification - Response
(Πρότασις – Επαγωγή - Συμπέρασμα*)

METHODS

INSPIRATION	PROCESSING	IMPLEMENTATION
Research	Thinking	Decision
Stochastic	Memory	Mobilization
Selection	Knowledge	Intervention
Harmonization	Archiving	Achievement

TOOLS

INFORMATION NOTIONS
ideas
thoughts - matrices- empirical ideas-
stimuli

* Aristotle, inductive, deductive reasoning

PLATE 12
HUMANS AND OTHER ANIMALS

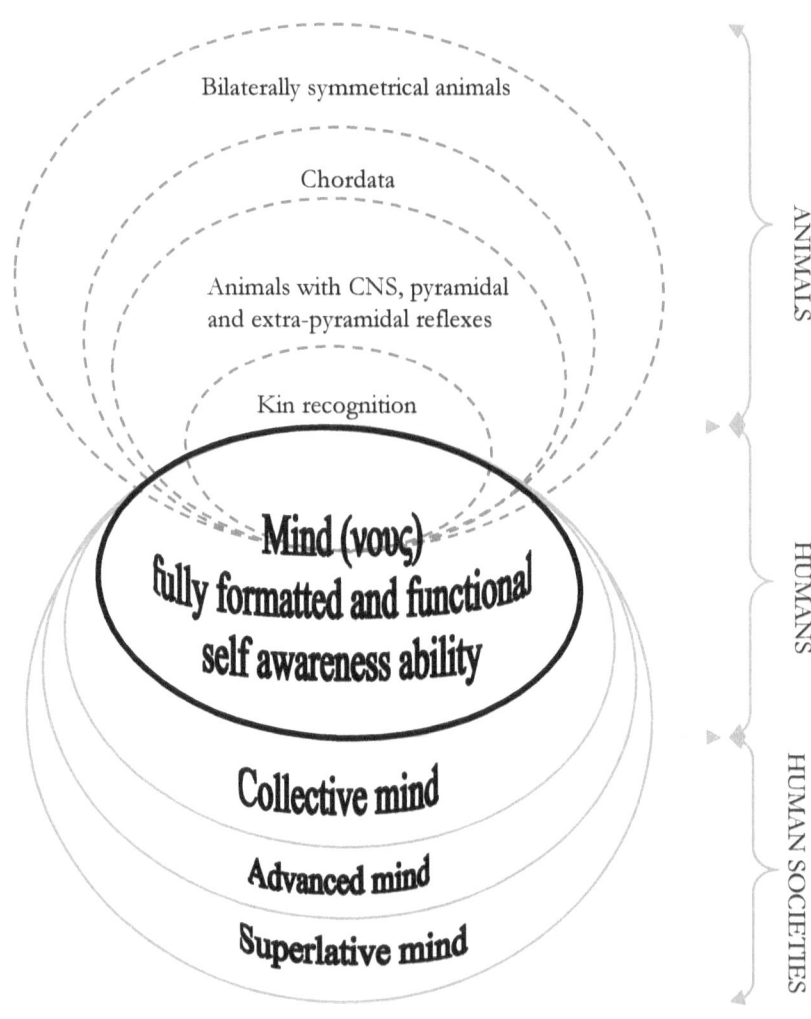

PLATE 13
MIND AND HUMAN BODY

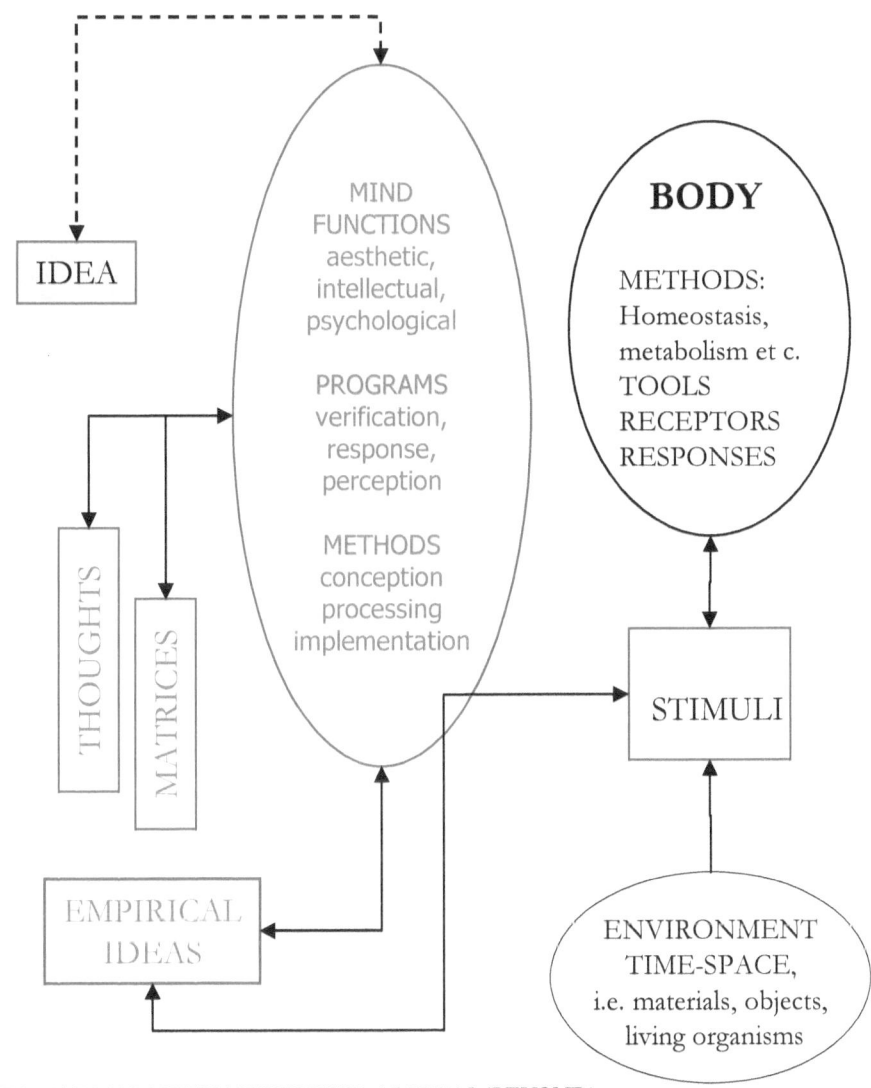

IMMATERIAL NOTIONS VERIFIED AS IDEAS (*BEYOND*)

MATERIAL (UNIVERSE) MATERIAL NOTIONS NOT VERIFIED TO IDEAS
WITHIN THE UNIVERSE THE IMMATERIAL NOTIONS (IDEAS, *LOGOS*)
CONSTITUTE THE *BEYOND*

PLATE 14
MIND, IDEA AND LOGOS

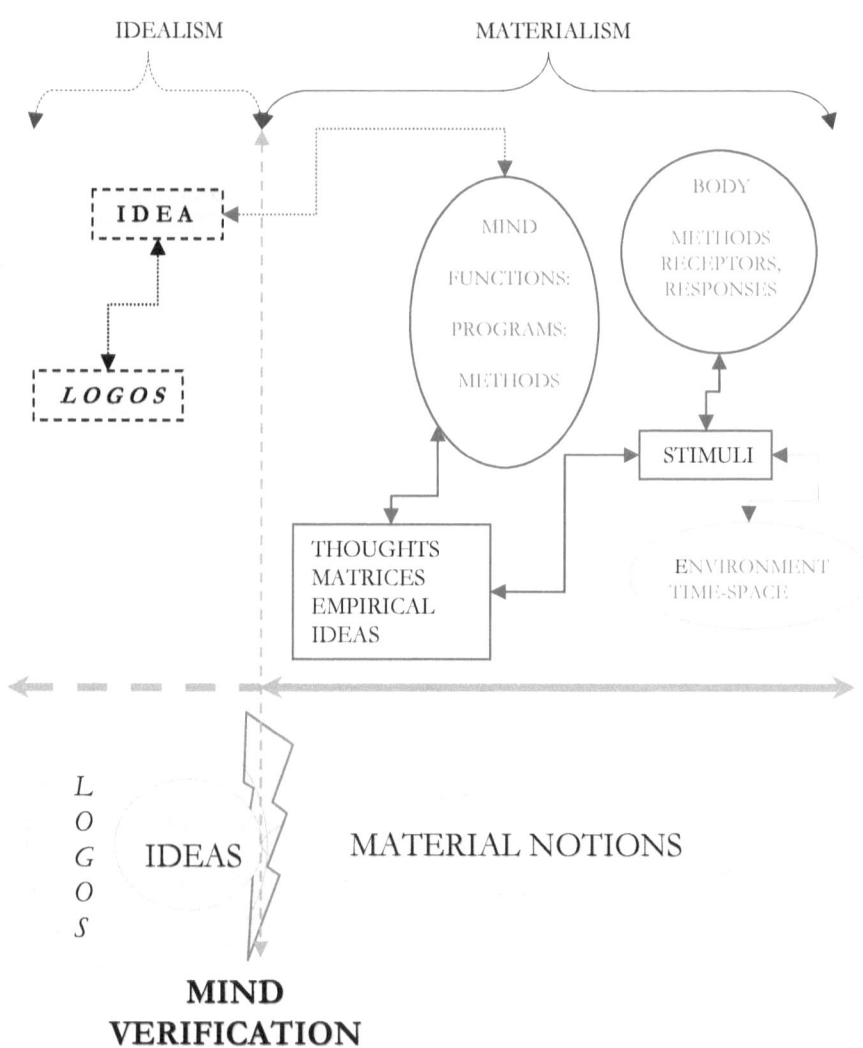

IDEALISM   MATERIALISM

IDEA

LOGOS

MIND
FUNCTIONS:
PROGRAMS:
METHODS

BODY
METHODS
RECEPTORS,
RESPONSES

STIMULI

THOUGHTS
MATRICES
EMPIRICAL
IDEAS

ENVIRONMENT
TIME-SPACE

L O G O S

IDEAS   MATERIAL NOTIONS

MIND
VERIFICATION

IMMATERIAL NOTIONS VERIFIED AS IDEAS (*BEYOND*)

MATERIAL (UNIVERSE) MATERIAL NOTIONS NOT VERIFIED TO IDEAS
WITHIN THE UNIVERSE THE IMMATERIAL NOTIONS (IDEAS, *LOGOS*)
CONSTITUTE THE *BEYOND*

PLATE 15
NOTIONS AND MINDS

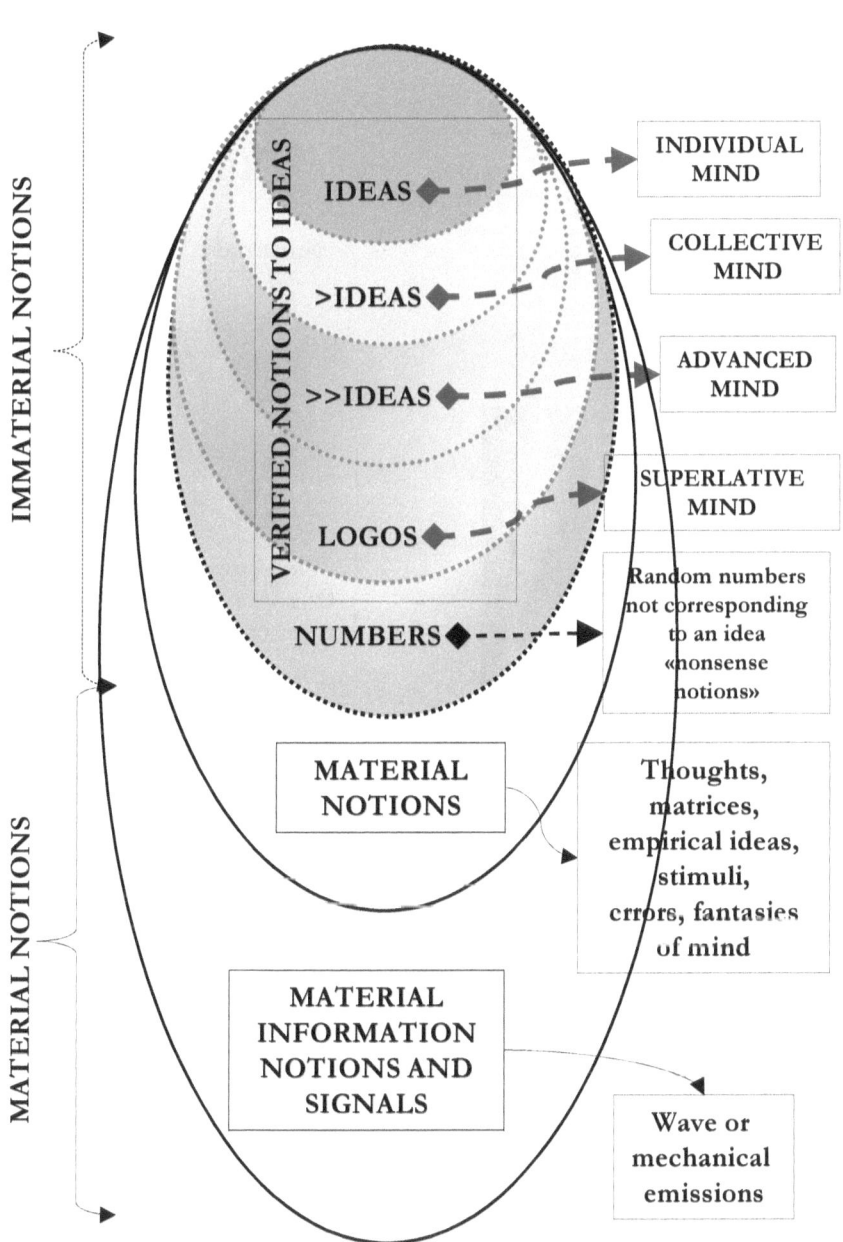

PLATE 16
INFORMATION PATHWAYS IN THE MIND

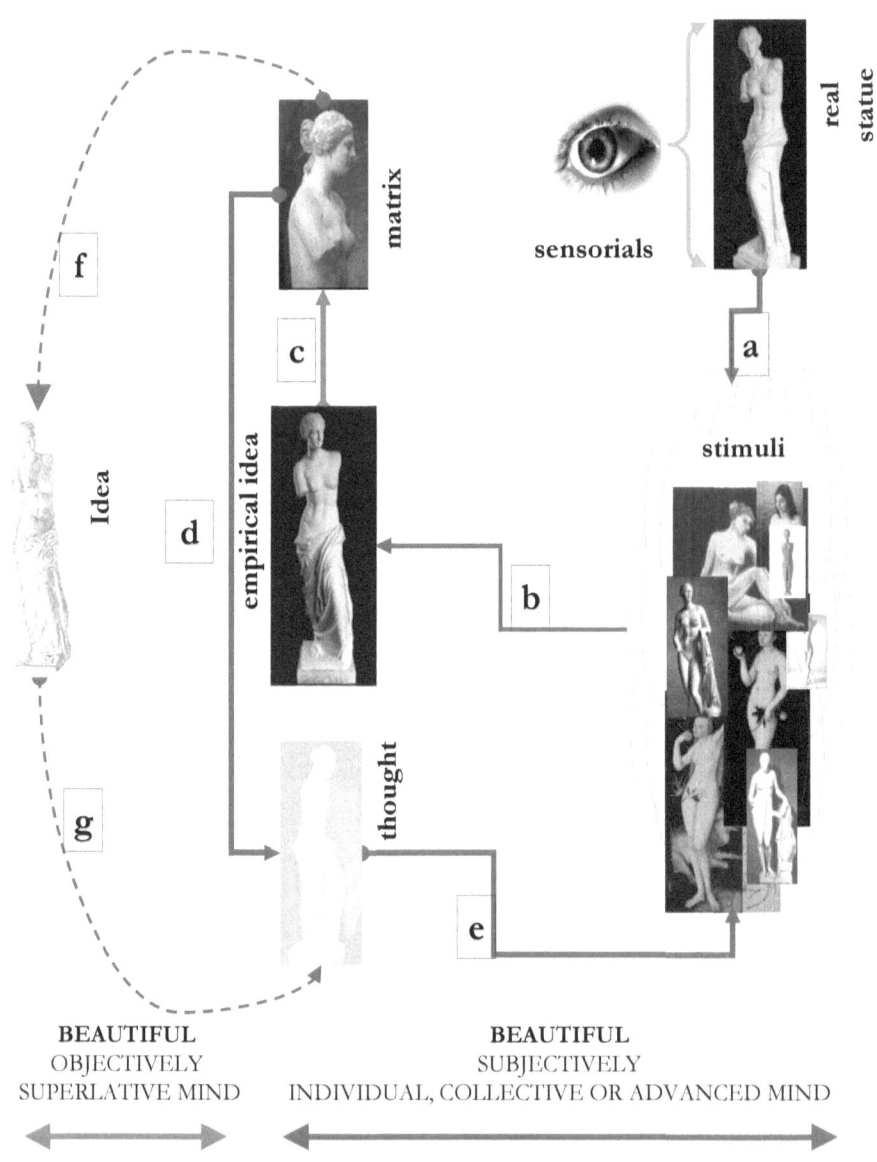

real statue

sensorials

stimuli

matrix

empirical idea

Idea

thought

c

d

f

g

a

b

e

BEAUTIFUL
OBJECTIVELY
SUPERLATIVE MIND

BEAUTIFUL
SUBJECTIVELY
INDIVIDUAL, COLLECTIVE OR ADVANCED MIND

PLATE 17
EVOLUTION PATHWAY

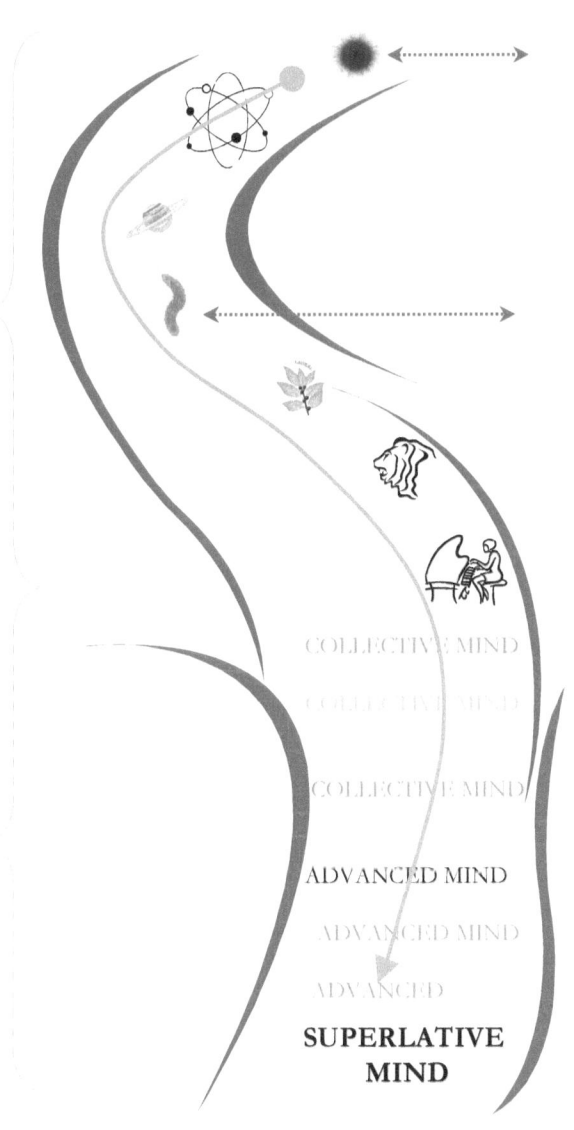

PRIOR TO HUMAN MIND

POST TO HUMAN MIND

1. BIG BANG

Elementary particles
Energy
Atoms
Molecules
Nucleotides
RNA

2. COMMON ANCESTRY ORGANISM

Microbes
Plants
Animals
Homo sapiens

3. INDIVIDUAL MIND

COLLECTIVE MIND

COLLECTIVE MIND

COLLECTIVE MIND

ADVANCED MIND

ADVANCED MIND

ADVANCED

SUPERLATIVE MIND

PLATE 18
MATERIAL - IMMATERIAL INTERCHANGE

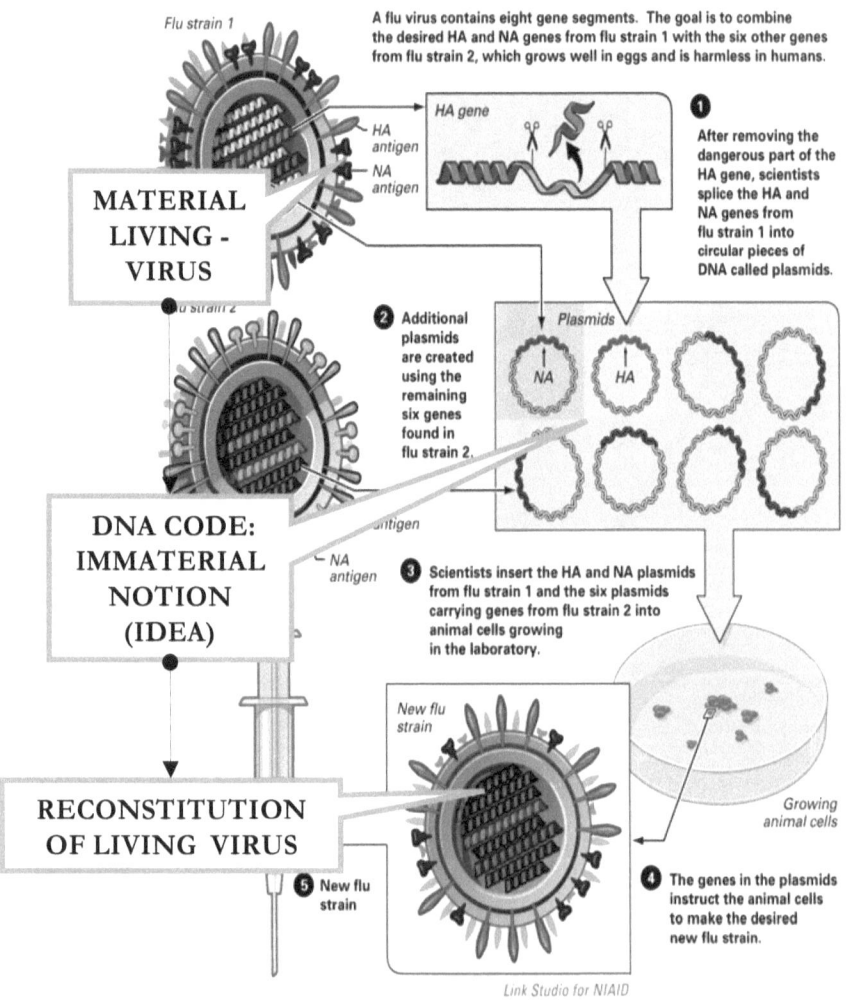

Flu strain 1

A flu virus contains eight gene segments. The goal is to combine the desired HA and NA genes from flu strain 1 with the six other genes from flu strain 2, which grows well in eggs and is harmless in humans.

HA antigen
NA antigen

MATERIAL LIVING - VIRUS

HA gene

1 After removing the dangerous part of the HA gene, scientists splice the HA and NA genes from flu strain 1 into circular pieces of DNA called plasmids.

u strain 2

2 Additional plasmids are created using the remaining six genes found in flu strain 2.

Plasmids

NA   HA

DNA CODE: IMMATERIAL NOTION (IDEA)

antigen

NA antigen

3 Scientists insert the HA and NA plasmids from flu strain 1 and the six plasmids carrying genes from flu strain 2 into animal cells growing in the laboratory.

New flu strain

RECONSTITUTION OF LIVING VIRUS

Growing animal cells

5 New flu strain

4 The genes in the plasmids instruct the animal cells to make the desired new flu strain.

Link Studio for NIAID

National Institute of Allergy and Infectious Diseases (NIAID)
National Institutes of Health (NIH)
Department of Health and Human Services (HHS)
http://www3.niaid.nih.gov/news/focuson/flu/illustrations/reassort_rev/
reversegenetics.htm

PLATE 19
MIND AND *BEYOND*
(NOYΣ KAI ΕΠΕΚΕΙΝΑ)

LEVELS	FUNCTIONS aesthetic, intellectual, psychological		INFORMATION (NOTIONS)
INDIVIDUAL MIND	VERIFICATION OF REAL (TRUE) AND SCARCE SCIENTIFIC NOTIONS	THE *BEYOND* (επέκεινα) — MATERIAL UNIVERSE	BODY STIMULI ↕ EMPIRICAL IDEA
COLLECTIVE MIND	VERIFICATION OF FEW SCIENTIFIC NOTIONS		MATRIX ↔ THOUGHT
ADVANCED MIND	VERIFICATION OF MOST SCIENTIFIC NOTIONS		
SUPERLATIVE MIND	VERIFICATION OF ALL NOTIONS INCLUDING AESTHETICS, ETHICS/ ACTIVITIES	IMMATERIAL THE *BEYOND*	IDEA ↕ *LOGOS* TOTALITY OF IDEAS

IMMATERIAL NOTIONS VERIFIED AS IDEAS (*BEYOND*)

MATERIAL (UNIVERSE) MATERIAL NOTIONS NOT VERIFIED TO IDEAS WITHIN THE UNIVERSE THE IMMATERIAL NOTIONS (IDEAS, *LOGOS*) CONSTITUTE THE *BEYOND*

PLATE 20
ASCENDING EVOLUTION OF MINDS

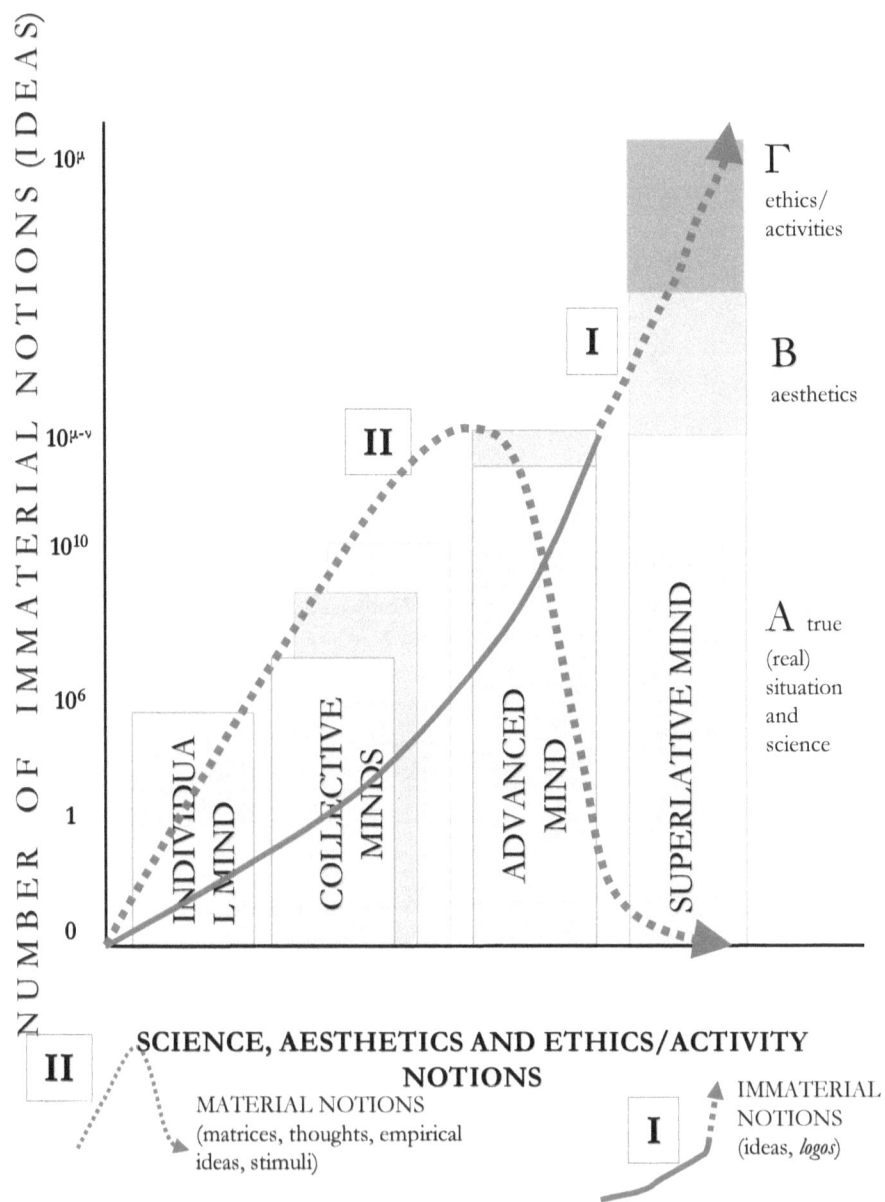

NUMBER OF IMMATERIAL NOTIONS (IDEAS)

10^μ

$10^{\mu-\nu}$

10^{10}

10^6

1

0

II

I

Γ
ethics/
activities

B
aesthetics

A true
(real)
situation
and
science

INDIVIDUAL MIND

COLLECTIVE MINDS

ADVANCED MIND

SUPERLATIVE MIND

**SCIENCE, AESTHETICS AND ETHICS/ACTIVITY
NOTIONS**

II

MATERIAL NOTIONS
(matrices, thoughts, empirical
ideas, stimuli)

I

IMMATERIAL
NOTIONS
(ideas, *logos*)

PLATE 21
NOTIONS CATEGORIES

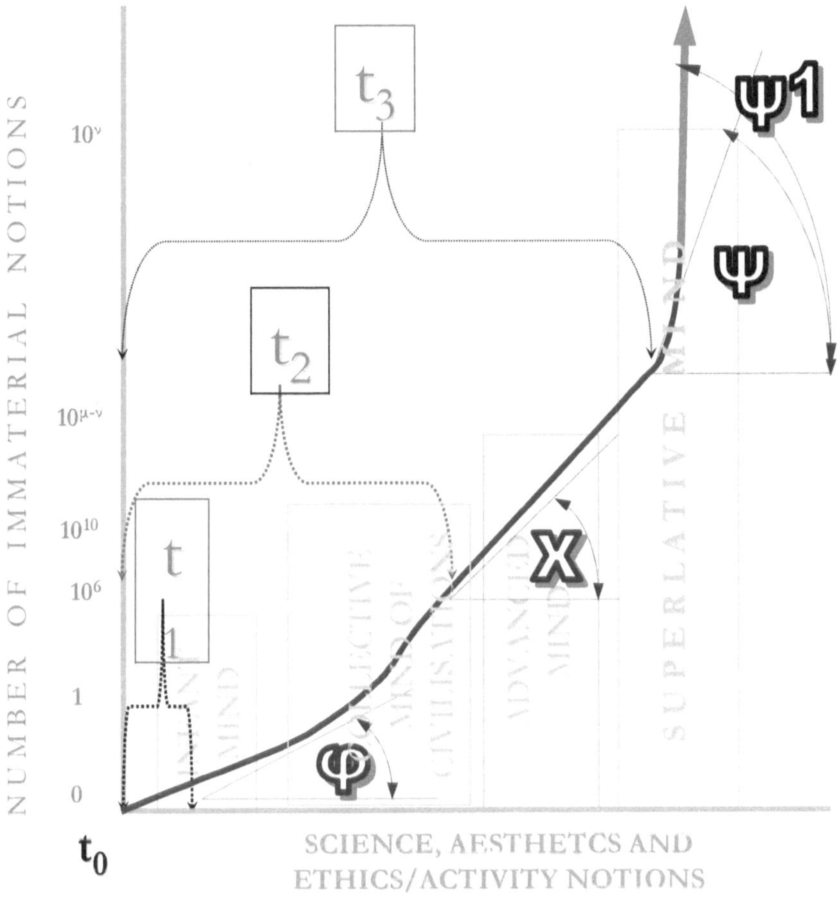

t_3

10^ν

Ψ1

Ψ

t_2

$10^{\mu-\nu}$

NUMBER OF IMMATERIAL NOTIONS

10^{10}

t

X

10^6

1

0

SUPERLATIVE MIND

t_0

SCIENCE, AESTHETCS AND
ETHICS/ACTIVITY NOTIONS

φ

t_0 Time that the first mind is functional

$t_{1,2,3}$ Time period till the number of immaterial notions is reached at each level of mind (individual, collective, advanced, superlative)

φ, χ, ψ angles of increase of notions between evolutionary points of different minds as to the scientific, aesthetics and ethics/activities

The angle ψ1 in the superlative mind is 90°

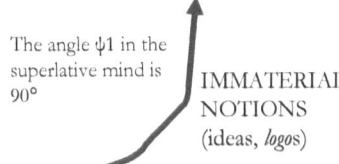

IMMATERIAL NOTIONS
(ideas, *logos*)

PLATE 22
EVOLUTION ASCENDING TREND

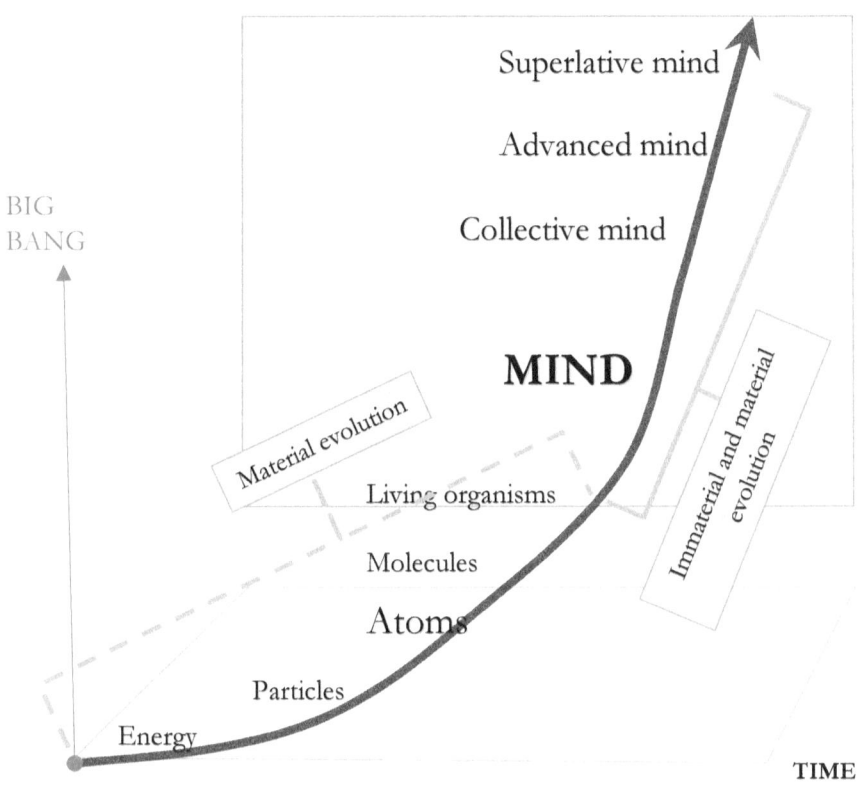

Superlative mind

Advanced mind

Collective mind

MIND

BIG
BANG

Material evolution

Living organisms

Molecules

Atoms

Immaterial and material evolution

Particles

Energy

TIME

Expansion of the universe with a parallel drop
of temperature and entropy

PLATE 23
HUMAN MIND CENTERED EVOLUTION

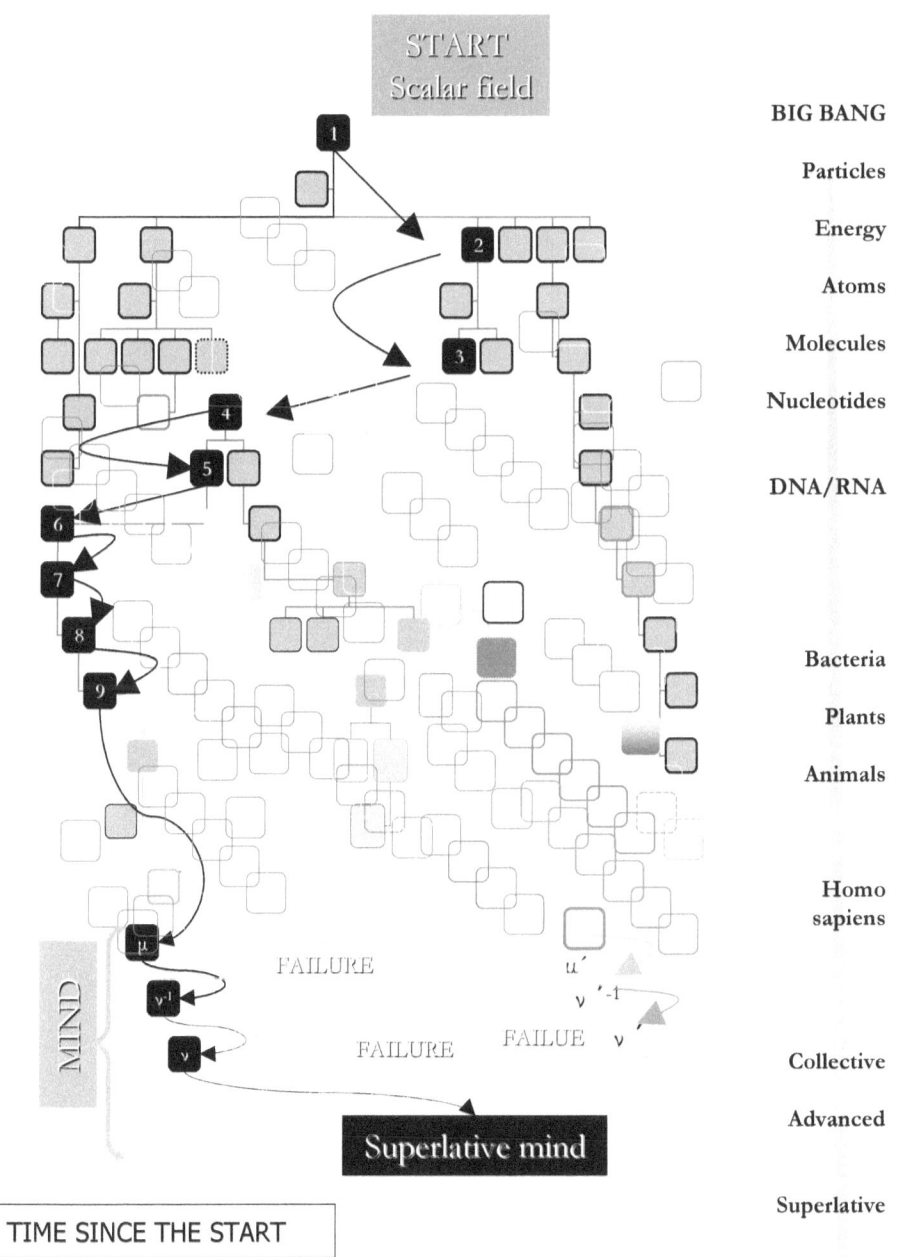

START
Scalar field

BIG BANG

Particles

Energy

Atoms

Molecules

Nucleotides

DNA/RNA

Bacteria

Plants

Animals

Homo
sapiens

Collective

Advanced

Superlative

FAILURE

FAILURE

FAILUE

MIND

Superlative mind

TIME SINCE THE START

PLATE 24
MIND

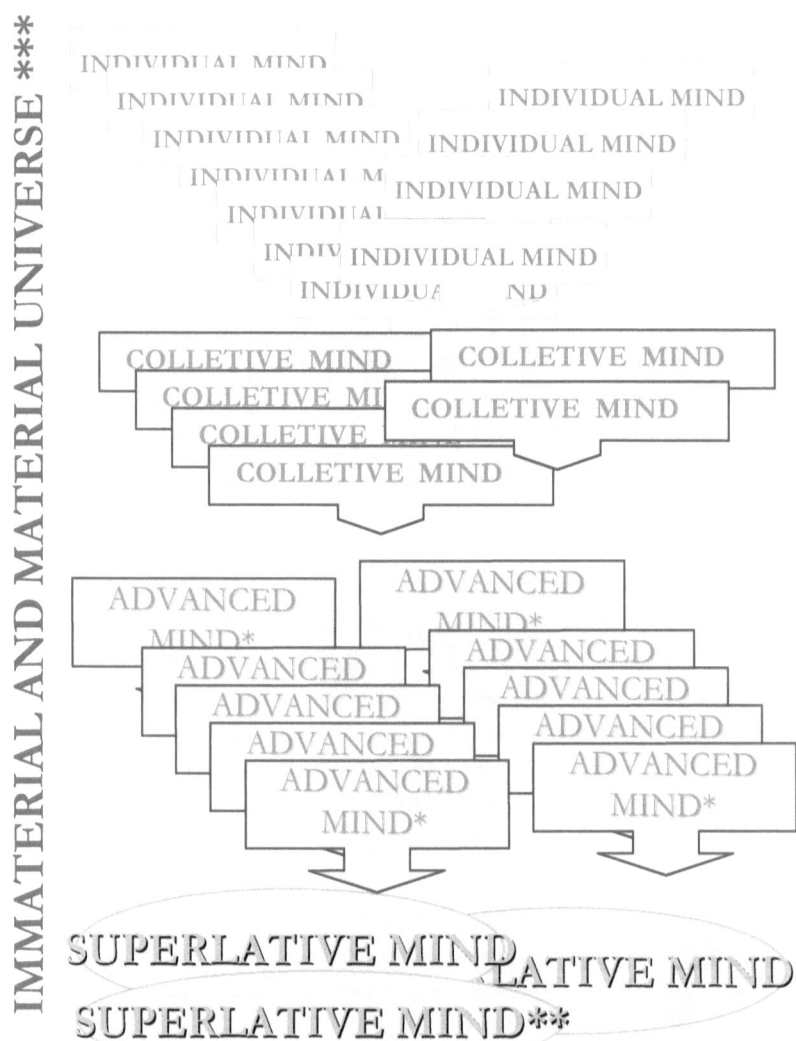

IMMATERIAL AND MATERIAL UNIVERSE ***

INDIVIDUAL MIND
INDIVIDUAL MIND
INDIVIDUAL MIND
INDIVIDUAL MIND
INDIVIDUAL M
INDIVIDUAL MIND
INDIVIDUAL
INDIV INDIVIDUAL MIND
INDIVIDUA ND

COLLETIVE MIND
COLLETIVE MIND
COLLETIVE MI
COLLETIVE MIND
COLLETIVE
COLLETIVE MIND

ADVANCED MIND*
ADVANCED MIND*
ADVANCED
ADVANCED
ADVANCED
ADVANCED
ADVANCED
ADVANCED
ADVANCED
ADVANCED
ADVANCED MIND*
ADVANCED MIND*

SUPERLATIVE MIND LATIVE MIND
SUPERLATIVE MIND**

*   Possibility of evolution of one or more advanced minds
**   Possibility in the universe or universes of pre existence or
       co-existence   or post-existence of superlative minds
*** Independent compatible co-existence of ideas *(logos)* and materials

PLATE 25
VIRTUALITY IN UNIVERSE AND *BEYOND*

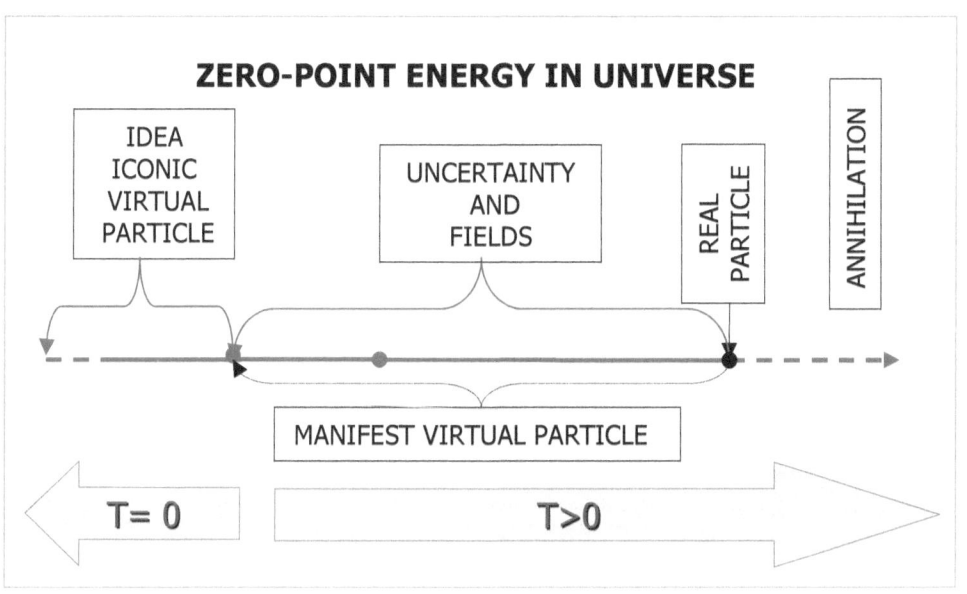

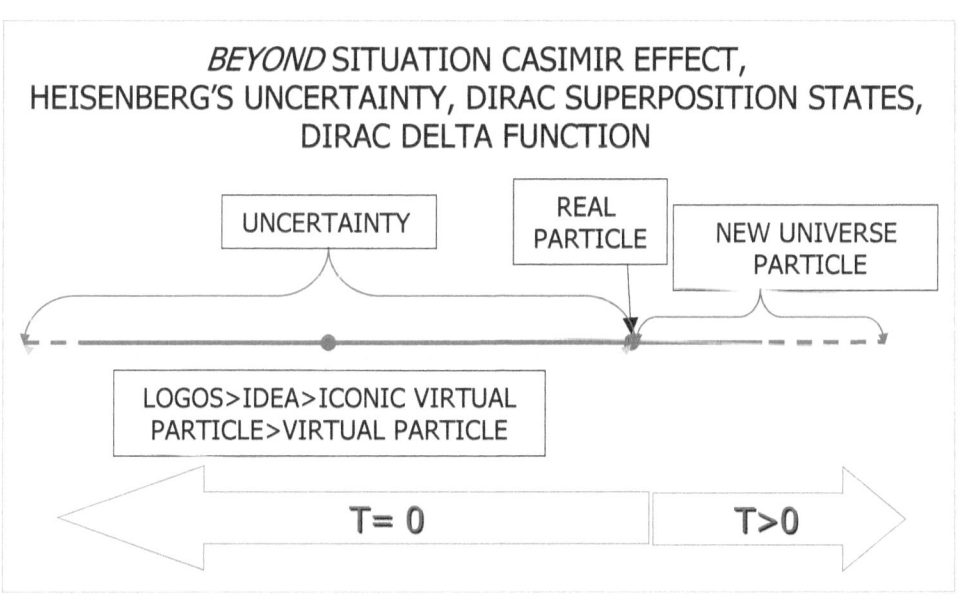

PLATE 26
BEYOND AND UNIVERSE

I. AUTOMATIC CREATION

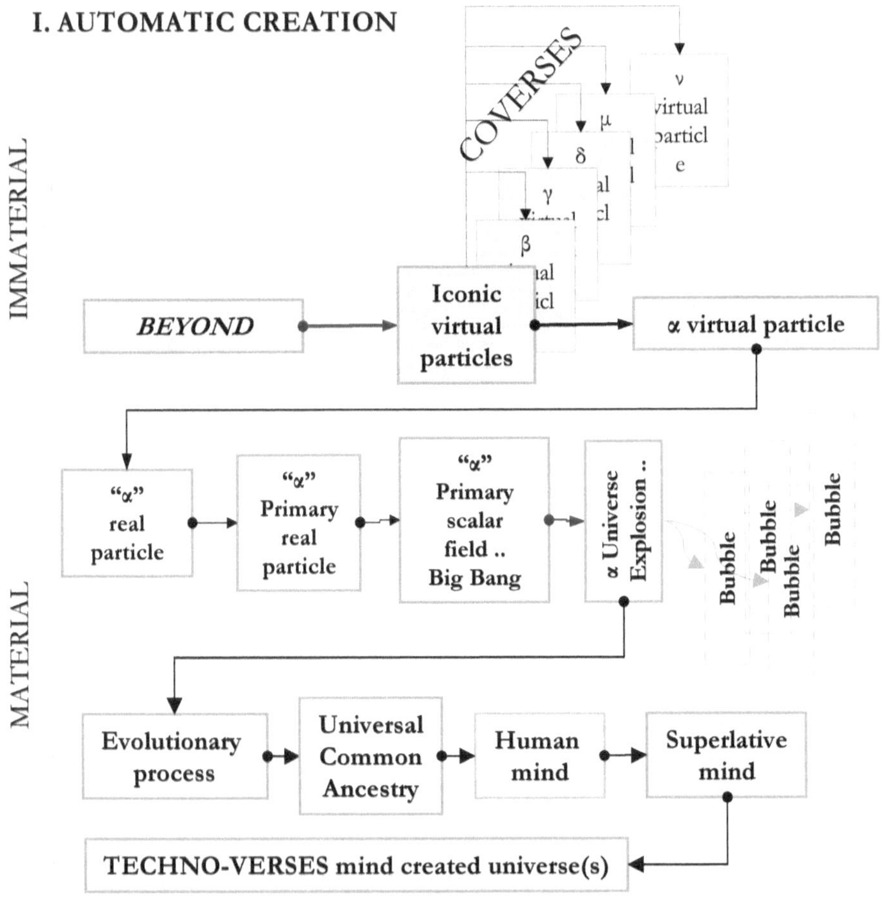

IMMATERIAL

COVERSES

ν virtual particle

μ

δ

γ

β

| **BEYOND** | → | Iconic virtual particles | → | α virtual particle |

MATERIAL

| "α" real particle | → | "α" Primary real particle | → | "α" Primary scalar field .. Big Bang | → | α Universe Explosion .. |

Bubble → Bubble → Bubble → Bubble → Bubble

| Evolutionary process | → | Universal Common Ancestry | → | Human mind | → | Superlative mind |

| **TECHNO-VERSES** mind created universe(s) | ← |

II. AN EXAMPLE OF A VIRTUAL PARTICLE AT THE ZERO POINT ENERGY IN THE UNIVERSE

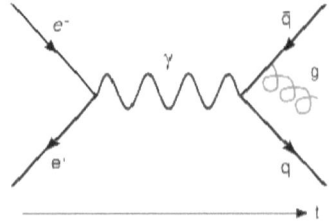

FEYMAN DIAGRAM

Two virtual fermions (virtual electron and virtual positron) swing from virtual to real (material) and collide producing a virtual photon γ that will become real and will collide in turn with two virtual quanta (positive and negative virtual quarks) that will be neutralized creating finally a real gluon.

PLATE 27
MIND-CENTERED RESTRICTIVE EVOLUTION

THE *BEYOND* AND THE ICONIC
VIRTUAL PARTICLE

PRIMARY REAL PARTICLE

PRIMARY SCALAR FIELD

BIG BANG

ENERGY

ELEMENTARY PARTICLES
ATOMS
MOLECULES
NUCLEOTIDES

DNA/RNA

UNIVERSAL COMMON
ANCESTRY
BACTERIA
PLANTS
ANIMALS
HOMO SAPIENS

HUMAN MIND

INDIVIDUAL

COLLECTIVE

ADVANCED

**SUPERLATIVE
MIND TECMOR**

TIME SINCE THE START (BIG BANG 1.3x10^{10} years ago)

PLATE 28
UNMOVABLE MOVER AND *BEYOND*

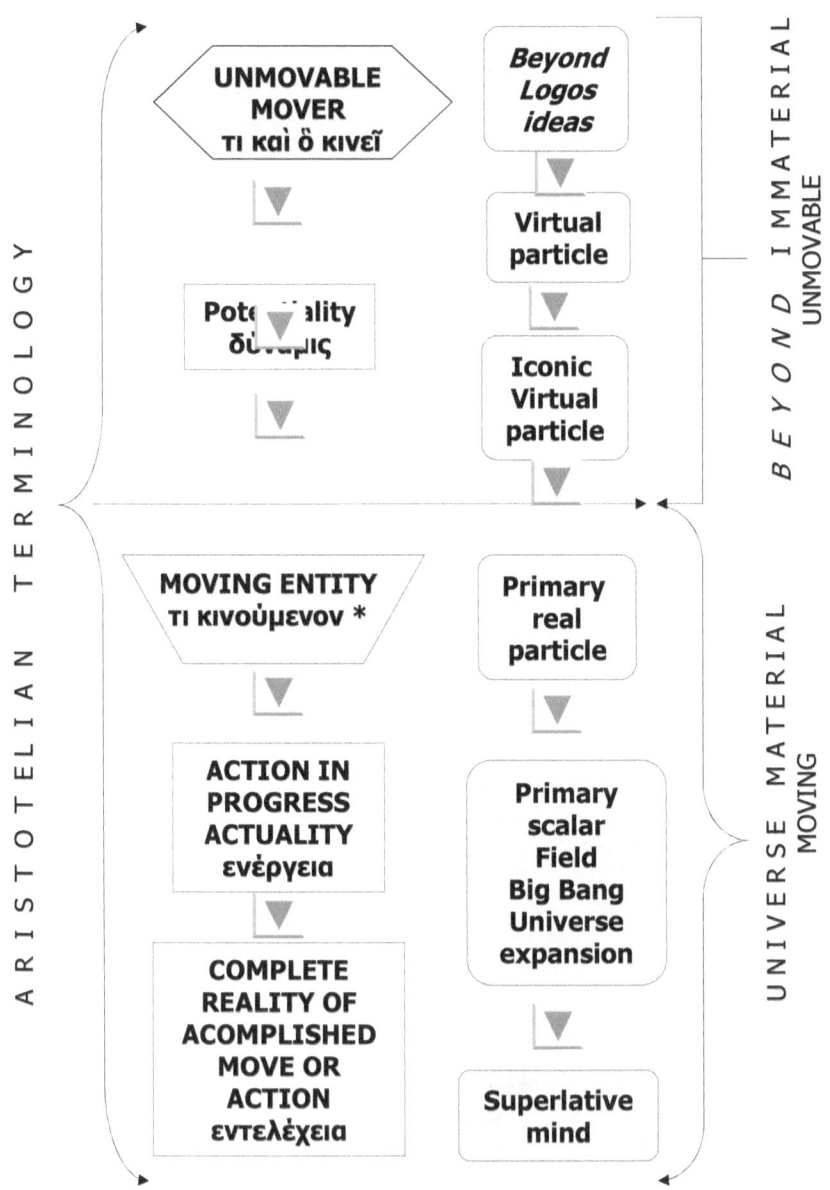

* ἐπεὶ δὲ τὸ κινούμενον καὶ κινοῦν [καὶ] μέσον

PLATE 29
GENERAL THEORY OF CREATION AND EVOLUTION

BEYOND

I D E A S
AS
INFORMATION NOTIONS

VIRTUAL PARTICLE

ICONIC VIRTUAL
PARTICLE

TIME
VARIABLE
EQUAL TO
ZERO
PRE BING
BANG

REAL PARTICLE

B I G   B A N G

$$\frac{F_c}{A} = -\frac{d}{da}\frac{\langle E \rangle}{A} = -\frac{\hbar c \pi^2}{240 a^4}$$

E N E R G Y   E X C E S S

I M P L O S I O N   O F   R E A L
PARTICLE

CREATIONAL PARTICLE THAT
WIILL EVOLVE TO A NEW
UNIVERSE

N E W   U N I V E R S E

PRIMARY SCALAR FIELD
BIG BANG

IMPLOSION FOLLOWED BY
EXPLOSIVE EXPANSION

ENERGY

ELEMENTARY PARTICLES

ATOMS

MOLECULES

NUCLEOTIDES

DNA/RNA

UNIVERSAL COMMON

HOMO SAPIENS

HUMAN MIND

RELATIVE TIME
MATERIAL PROCESS
VALIDITY OF E= mc^2